The Study of
Animal Bones from
Archaeological Sites

International Series of Monographs on Science in Archaeology

Consulting editor G. W. DIMBLEBY

Forthcoming titles in the Series

Carbon-14 Dating
R. BURLEIGH

Methods of Physical Examination in Archaeology
M. S. TITE

Soil Science in Archaeology
S. LIMBREY

Skins, Leathers and Parchments in Archaeology
R. REED

Surveying Methods in Archaeology
R. E. LININGTON

The Study of Animal Bones from Archaeological Sites

RAYMOND E. CHAPLIN
The Anatomy School, Cambridge
England

1971

SEMINAR PRESS · London and New York

SEMINAR PRESS LTD.
Berkeley Square House
Berkeley Square
London WIX 6BA

U.S. Edition published by
SEMINAR PRESS INC.
111 Fifth Avenue,
New York, New York 10003

Copyright © 1971 By SEMINAR PRESS LTD.

All Rights Reserved
No part of this book may be reproduced in any form
by photostat, microfilm, or any other means, without
written permission from the publishers

Library of Congress Catalog Card Number: 77–162377
ISBN: 0–12–816050–0

Printed in Great Britain by
William Clowes & Sons, Limited
London, Beccles and Colchester

Preface

In this book I have endeavoured to cover the more important theoretical and practical aspects of bone studies in archaeology and show how certain basic principles can be applied to sites of any age. I have sought to illustrate the validity and unity of a socio-biological approach by emphasising both the importance of the human factor in bone accumulations, and the ecology and biology of the animals as a permissive force in human advancement.

R.E.C.

Cambridge, 1971

Acknowledgements

I am very conscious of the contribution of friends, colleagues, and students to the work that has led to this book and I hope that it will, in part, repay the debt I owe them. My special thanks go to Jill Chaplin, Jennie Coy, Dorothy Nicholas, the late Kenneth Marshall, Professor Ian Silver, Dr Ian Cornwall, and Professor Geoffrey Dimbleby, all of whom gave freely of their skills and knowledge in the preparation of the manuscript. The excellent photographs were prepared by Mr T. Crane and staff of the Audio-Visual aids department of the Anatomy School, Cambridge and the radiographs by Mr J. Fozzard. Dr Donald Thomas kindly loaned the slides from which Fig. 7 was prepared. Figures 25 and 26 are reproduced here by courtesy of the Birmingham Archaeological Society.

R.E.C.

Contents

1
Structure and Biological Properties of Bone — 1

The skeleton and its structure 1
Development and growth of bone 7
Teeth 12
Physical and chemical properties of bone 12
Factors influencing the survival of buried bone . . . 13

2
Planning and Organisation of Bone Studies — 20

Bone studies in archaeology 20
Requirements for collecting bones in the field . . . 24
Whole skeletons 29
Labelling, packaging and transporting bones . . . 31
Preliminary tasks to be carried out by the specialist . . 35

3
Bone Identification and the Establishment of Reference Collections — 37

Problems of identification 37
Comparative osteological collections 41
The specialised osteological collection 49
The preparation of comparative material 50

4
Techniques for the Study of Site Collections — 55

Planning of the site study	55
The cultural context	55
Topic definition and the examinations to be made	58
Working procedure	63
Quantifying the species	63
fragments method	64
weight method	67
minimum numbers method	69
Determination of the minimum number of animals	70

5
Age Determination from Bones — 76

Tooth eruption	78
Epiphyseal fusion	80
Suture closure	81
Incremental structures	84
Tooth wear	85
Antler development	89
Growth	89
Qualitative features	90

6
The Use of Measurements — 91

Measuring a bone	95
Measuring technique and accuracy	98

7
The Determination of Sex from Bones — 100

8
Bone Pathology — 108

Abnormalities of development 109
Metabolic disorders of bone 110
Hormonal influences 112
Necrosis and inflammation of bones 112
Neoplastic and similar conditions 113
Bone discontinuities 114
Diseases of the joints 114

9
The Interpretation of Bone Evidence — 120

Limiting factors of the archaeological evidence . . . 120
The use of ancillary evidence 122
Site examples 123
    Site 1. A Saxon farm, Whitehall, London . . . 124
    Site 2. Medieval workshop debris 138

10
Animal Remains as Indicators of Past Environments — 143

Terrestrial mammals 143
Marine mammals 155
Reptiles, amphibians and fish 155
Birds 155

References — 160

Subject Index — 165

1
Structure and Biological Properties of Bone

The Skeleton and its Structure

The archaeologist is dealing with a bone that is long dead, the physical properties and chemical composition of which differ from living bone. Although we are concerned largely with the morphology of the bone which changes little with burial it is necessary to appreciate the biological characteristics and properties of bone in the living animal, and the way in which the bone is formed in the foetus and modelled throughout life. In the living animal the bones are complex living structures which perform many functions and which are capable of some modification in response to the needs of the animal whether physiological, as during lactation, or physical, in response to trauma.

In the vertebrates individual bones form an integrated structure —the skeleton—which in turn is an integral part of the whole body. This skelton lies in and about the soft tissues of the body and being entirely internal in the higher vertebrates it is termed an endo-skeleton. This internal skeleton contrasts with the external or exo-skeleton found in many invertebrates such as insects and molluscs. Some lower vertebrates possess both types of skeleton as in the fish where there may be an internal hard skeleton as well as an external one composed of scales or plates. This book is, however, concerned with the internal skeleton.

The semi-rigid framework of the vertebrate skeleton has several functions. It has the strength to support the weight of the soft tissues and additionally is a mechanical system providing attachments for the contractile and elastic tissues such as muscle, ligament and tendon enabling the animal to move about. It also has a protective function forming a rigid structure enclosing vital soft tissues such as the brain, heart and lungs. In addition to these mechanical functions bones also act as organs which can store and release minerals such as calcium and phosphate, and in the long bones there is a central cavity which may contain red marrow, which produces red corpuscles.

Although in future chapters when we consider dead bone we shall be talking almost exclusively of the mineralised structure of the bone it will be appreciated that as living tissue bones are endowed with blood vessels, nerves and enveloping membranes and are influenced by hormones. It is these soft tissue components which enable the bone to function also as a physiological element of the body and not solely as a mechanical structure.

The bones of the skeleton differ in their form, internal construction and function (Figs 1 and 2). Morphologically we can distinguish between the elongated cylindrical bones commonly termed long bones, which are confined to the limbs. Their mechanical function lies in supporting the body and providing a leverage system for the muscles and tendons. Contrasting with these are the flattened bones of the cranium, ribs and pelvis. These flattened bones have two functions: they protect underlying tissue such as the brain and their large surface area is important for the attachment of muscles.

Compact bones such as the carpals and tarsals have a very specific function and occur at particular sites. They are in general small and sub-rectangular in form (Fig. 1). They are important in the dissipation of stresses, especially compression; hence their occurrence in the wrist and ankle joint. They also provide a small but effective attachment for tendons at complex sites, enabling forces exerted by these tendons to change directions.

Vertebrae are a class of their own since they combine elements from the other broad groups. Thus the central neural canal containing the spinal cord is protected by the neural arch of each vertebra, the centrum is a compact body resisting compression and providing a rigid girder while the flattened spines and wings provide areas for the attachments of muscles, tendons and ligaments (Fig. 1).

The structure of long bones is best studied from a longitudinal section. Functional characteristics are best seen in a section that has been cut fresh and fixed to preserve the organic matter. The mechanical structures are, however, best appreciated in bones that

1. STRUCTURE AND BIOLOGICAL PROPERTIES OF BONE

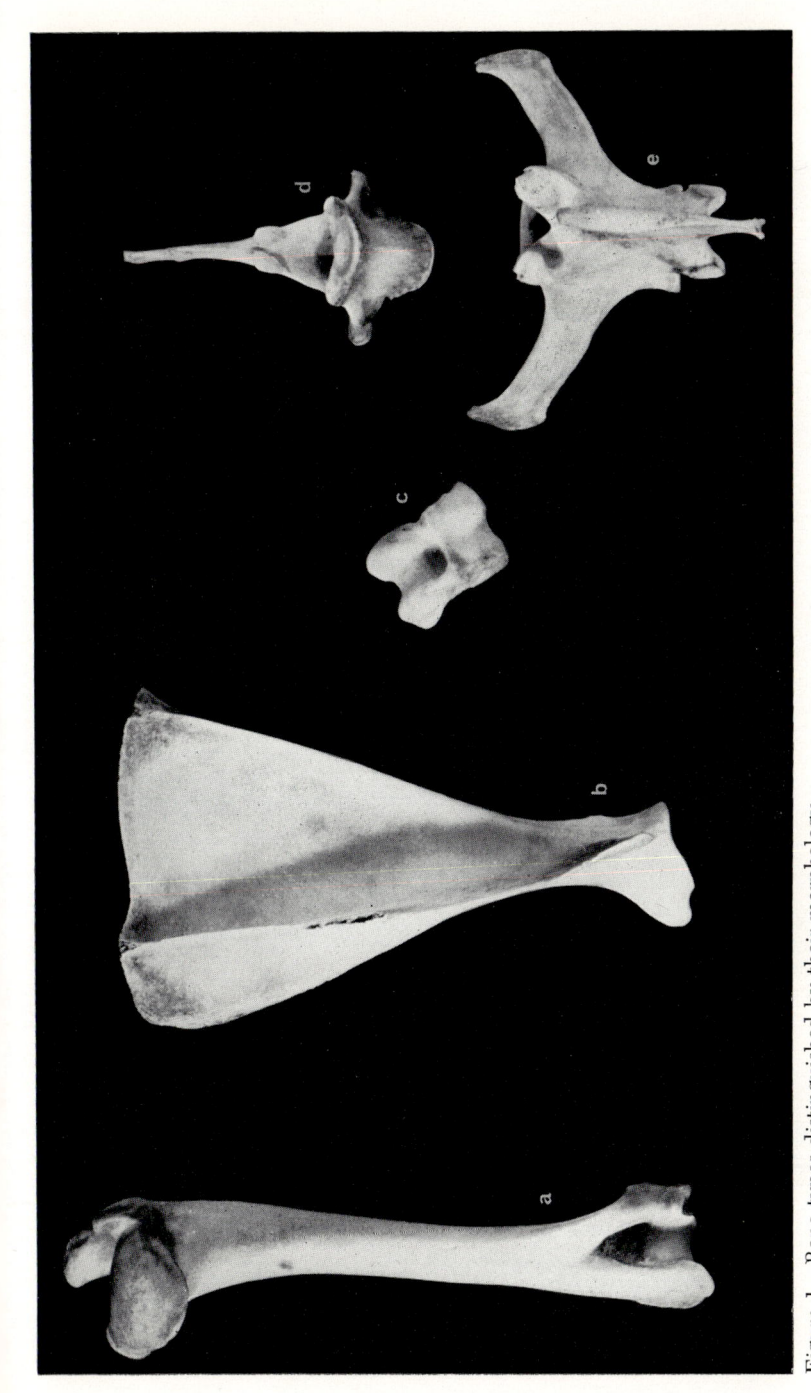

Figure 1 Bone types distinguished by their morphology.
(a) Long bone; humerus. (b) Flattened bone; scapula. (c) Compact bone; astragalus. (d) Thoracic vertebra. (e) Lumbar vertebra. Vertebrae embody elements of several types.

4   THE STUDY OF BONES FROM ARCHAEOLOGICAL SITES

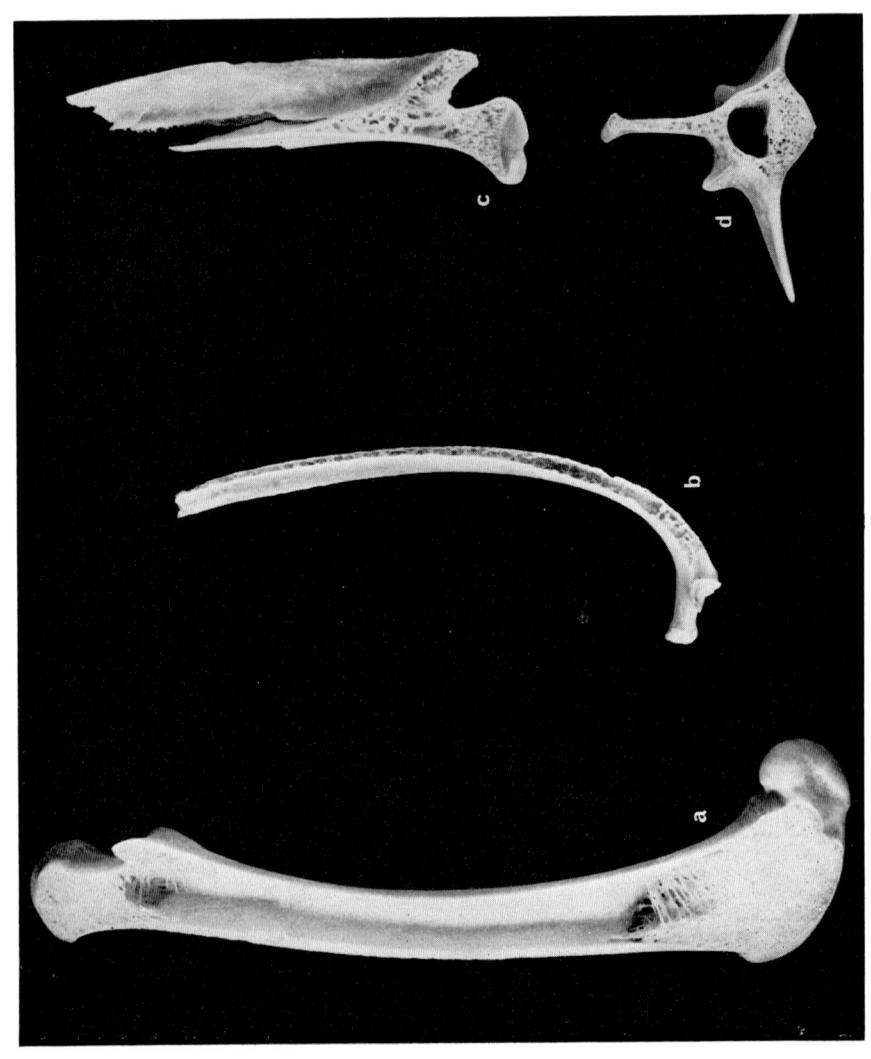

Figure 2   The internal structure of the different types of bones.
(a) Longitudinal section of a long bone: femur. (b) Longitudinal section of a flattened

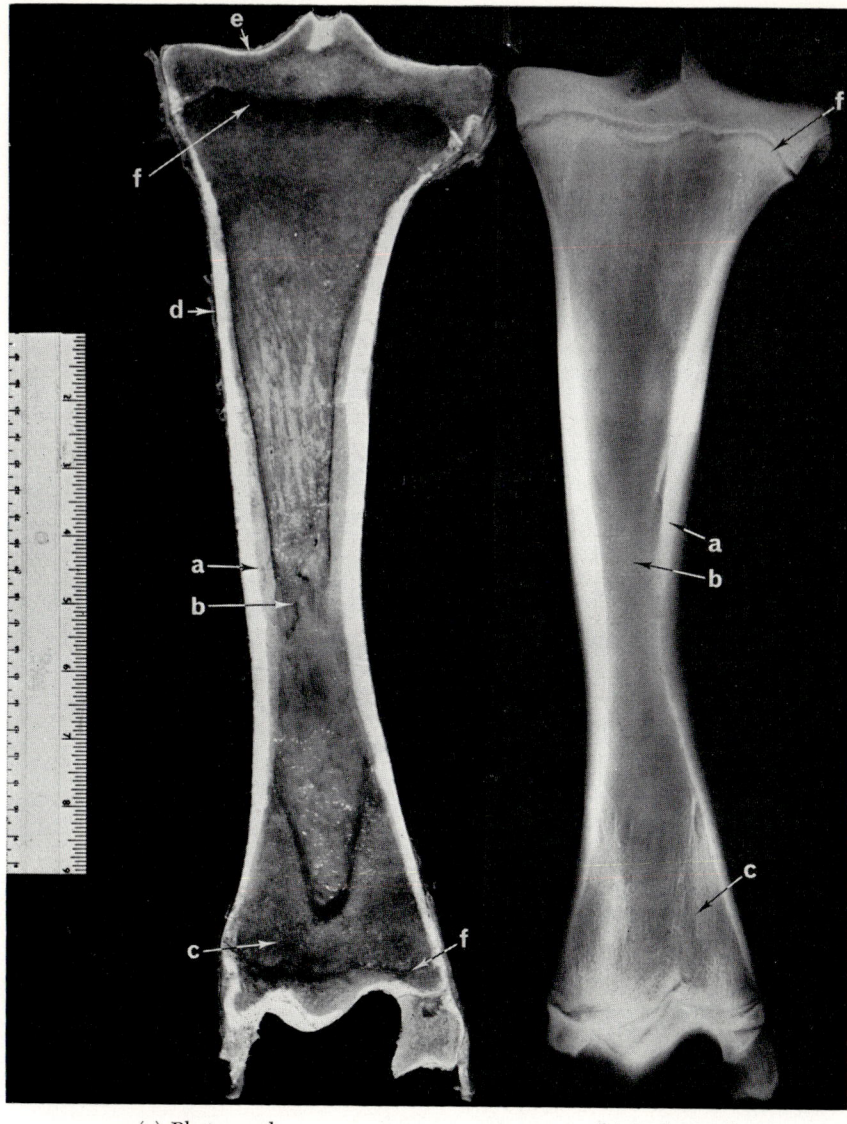

(a) Photograph (b) Radiograph

Figure 3 Longitudinal section of a fresh immature bovine tibia illustrating the relationship between hard and soft tissues.

(a) compact bone. (b) marrow and fat occupying the medullary cavity. (c) the network of cancellous bone in the distal end of the shaft. (d) teased portion of the periosteum. (e) cartilage over the articular surface. (f) cartilage separating the epiphyses from the shaft.

have been processed to remove the external organic material. Alternatively, structure can be studied from radiographs.

Examination of the inorganic section (Fig. 3a) shows that a long bone is composed of an outer frame of compact bone enclosing a central cavity, the extremities of which are filled with a sponge-like network of bone. The outer wall of compact bone varies in thickness. Maximum thickness is found about the middle of the length and tapers off towards the extremities and is extremely thin and compact over the joint surfaces. The cellular structure of the extremities is regular and the pattern is a result of the stresses borne by that bone (Fig. 3b). Thus consideration of the loading of each bone permits the recognition of plates of tension and compression within the cellular bone. In the preserved bone it will be seen that the central or medullary cavity and the interstices of the cellular bone are filled by marrow or in adult bones by fat. The marrow cavity is lined by an osteogenic (bone forming) membrane, the endosteum. Externally the bone, except over the articular surfaces, is covered by a similar membrane, the periosteum. The joint surfaces are covered by cartilage. Close examination will reveal the course of the blood vessels and nerves both internally and externally. Particular note should be made of the mode and place of entry of these in order to appreciate in dead bone the significance of the minute holes or foramina through which they enter.

The compact bones of the wrist (carpal) and ankle (tarsal) joints are of a broadly similar construction to the ends of the long bones. Internally they are composed of sponge-like or trabeculated bone and are covered externally by compact bone (Fig. 4). The arrangement of the internal plates again reflects the direction and magnitude of the forces to which they are subjected.

The form of the plate-like bones varies according to their site and function and this is reflected in their structure. A rib in cross section (Fig. 2) is seen to consist of an outer layer of compact bone filled internally with trabeculated bone. In the scapula the thicker areas are composed of outer layers of compact bone supported by trabeculated bone (Fig. 2). In the very thin areas the trabeculated bone is absent and the two compact layers are united.

The amount of compact and trabeculated bone are determined by function and load, the arrangement of the bone providing maximum strength with a minimum of weight. As function can vary during life, as also can load, a bone must be able to adjust itself to changing demands. Extra loading results in the deposition of new bone at the site of stress whilst disuse and inactivity, or conditions of weightlessness result in the resorption of bone. In addition to changes resulting from varying mechanical demand changes also occur in relation to physiological processes. Thus during pregnancy and lactation mineral may be removed from the bones whilst at other times it is deposited.

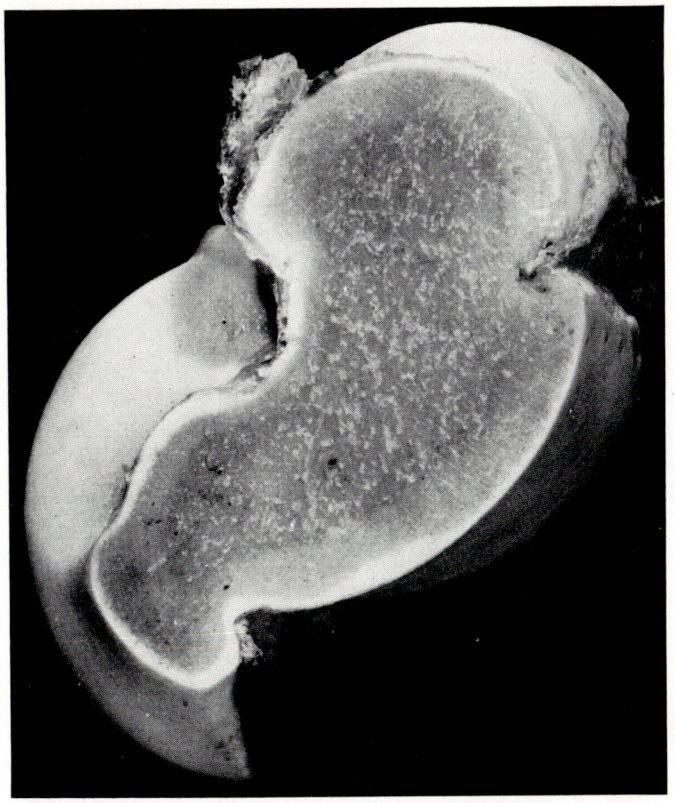

Figure 4 Longitudinal mid-line section of a fresh bovine astragalus showing the structure of a compact bone.

Development and Growth of Bone

In the foetus there are two precursors of bone, cartilage and fibrous tissue. The process of bone formation or osteogenesis in these tissues is carried out by the osteoblast cells. The bones which have their origin in the fibrous tissue (which includes the majority of the bones in the cranium) are generally of tabular form. Because of their origin in the fibrous sheets these are termed membrane bones. Those bones which have cartilaginous tissue as their precursor in the foetus are termed cartilage bones. These comprise the majority of the long and compact bones of the body.

As a result of their differing origins it will be apparent that two ossifying (bone forming) processes are involved. Thus ossification of the fibrous tissue to form membrane bones is referred to as intramembranous ossification whilst that of cartilage is termed endochondral ossification.

The formation of a membrane bone in the foetus is effected by ossification of the fibrous tissue. Ossification proceeds from a central point towards the periphery building up a network of ossified fibres or platelets which gradually coalesce and thicken. The outer layer of the original tissue becomes the periosteum and from this the bone is built up until it reaches the definitive thickness. Growth continues during ossification and frequently continues into postnatal life. Growth proceeds until the definitive form and size are achieved. This is possible because until the final size is reached adjacent bones are separated from one another by growing fibrous tissue. Thus it is cessation of growth by the fibrous tissue which gives these bones their form and size, the osteoblast cells converting the fibre into bone. The development of these bones is best seen in the skull.

The process of formation of a long bone by endochondral ossification is slightly different. The foetal skeleton of cartilage is covered by a membrane, the perichondrium, which is the primitive periosteum. Osteoblast cells migrate from the superficial perichondrium into the cartilage where they calcify the matrix or ground substance. Blood vessels extend into the calcifying zone and the cartilage cells shrink and then disappear leaving small marrow cavities which are partly occupied by filament-like processes of the osteogenic tissue. There is thus formed a complex structural network of bony platelets. This primitive structure is the basis on which the osteoblast cells will build the definitive bone. From outside bone formation also proceeds from the perichondrium. The calcified cartilage is now broken down and absorbed by osteoclast cells and replaced with bone deposited by the osteoblast cells. At this stage the interior of the bone consists of the trabeculae of the primitive spongy bone. In the central area this is resorbed to produce the marrow or medullary cavity whilst it remains in the extremities throughout life.

These processes are active throughout foetal and postnatal life until the definitive form and size are achieved. Thereafter the same processes produce the minor changes in structure necessitated by the stresses of use.

A long bone has three primary centres of ossification: the shaft or diaphysis and the two extremities or epiphyses. In many long bones there are secondary centres of ossification which provide for the development of individual processes such as the third trochanter of the equine femur. The foregoing description explains the process of growth apart from that of increasing length. Longitudinal growth is achieved by means of the cartilage which separates the ossification centres (Figs 3a and b). Thus during growth the shaft of the bone is separated from the ends by unossified growing cartilage.

In longitudinal growth the cartilage is growing at one face whilst ossification proceeds at its base. During active longitudinal growth the formation of new cartilage proceeds at the same rate as ossification. When cartilaginous growth ceases, the whole becomes ossified and the diaphysis and epiphysis are united and no further longitudinal growth can occur. Often, however, the epiphysis of one extremity unites before the other so that elongation now continues at a single point until the definitive size is attained and this also unites with the shaft (Fig. 5). Longitudinal growth is then complete and no further elongation will occur. The fusion of the epiphyses proceeds in an orderly fashion and a knowledge of the age at which an epiphysis unites with the shaft is one of the most important methods of estimating the age at death from animal bones found on excavations. This method of age estimation is also widely used in forensic medicine and in the verification of the age of sub-adult people in certain circumstances.

Rather surprisingly most textbooks describe the structure of bone as an inner and outer layer separated by a Haversian layer. Whilst in part this is typical of some human bones and those of a few other species it is not a typical arrangement in mammals as a whole. The varieties of bone structure that occur in a single bone are due to the remodelling process that is an integral part of longitudinal growth and the formation of a bone's definitive shape. This growth process and the structure and origin of the types of bone are discussed by Enlow and Brown (1956–58) and Enlow (1963). The reader is strongly recommended to consult these works.

The overall growth of a long bone is completed when both

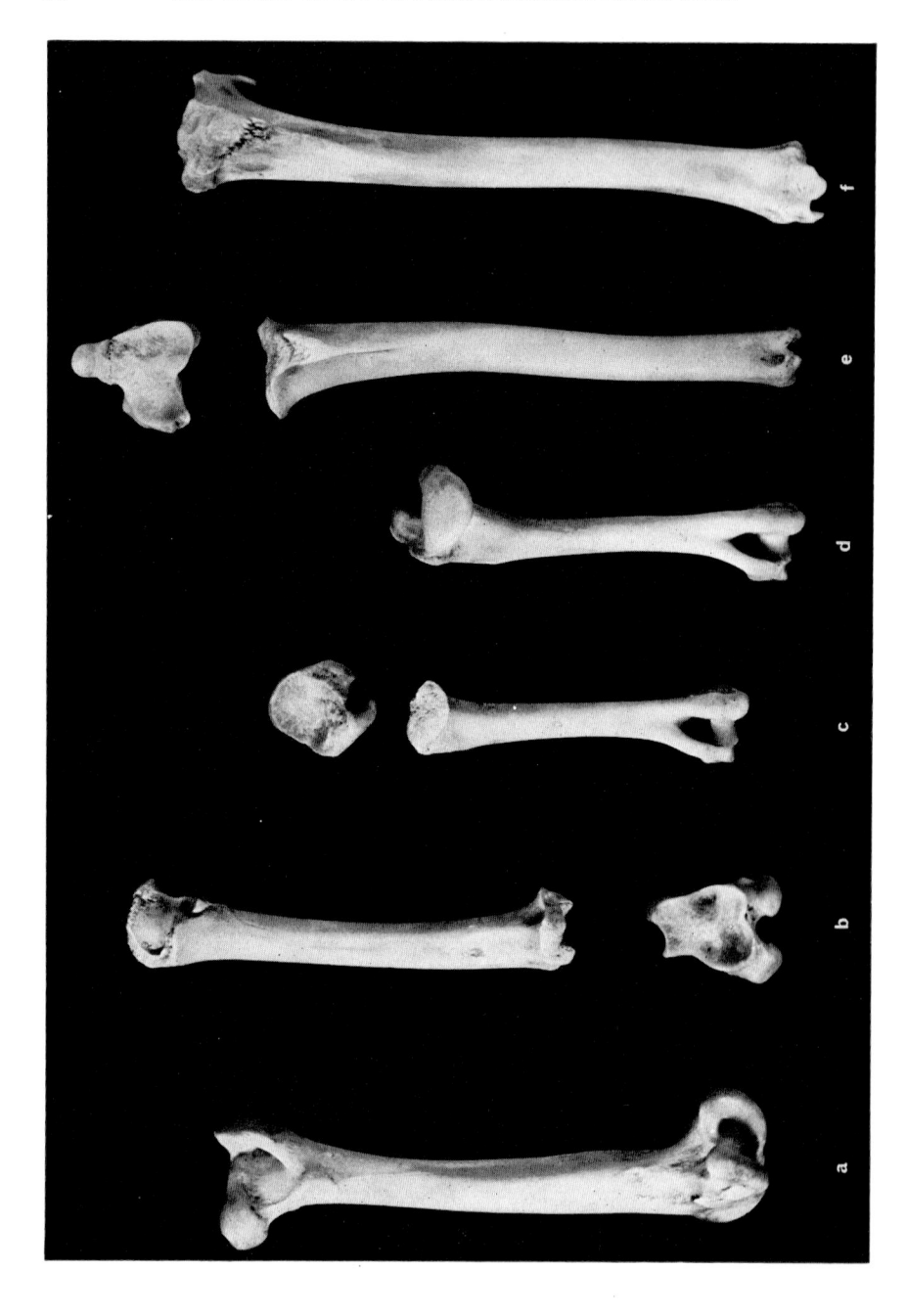

10 THE STUDY OF BONES FROM ARCHAEOLOGICAL SITES

epiphyses have united with the shaft. At this same time the osteogenic layer of the periosteum becomes much reduced. Comparison of the same bone between an old and a recently matured animal does, however, reveal many minor points of difference. The surface of the younger bone is much smoother and lacks many of the roughenings and ridges found on the older one. These secondary sculptings result from the action of muscles and tendons and are formed by the periosteum in response to the forces exerted by the elastic tissues. In old age tendons and ligaments may also become partly ossified about their attachments.

Injury to the periosteum such as a break or fracture of a bone or even pressure results in the renewal of the rapid osteogenic processes. Thus fractured and broken bones are repaired by a process of very rapid but often irregular deposition of bone. This repair does not remodel the bone or even rebuild it to the basic plan of the undamaged bone. Rather it buttresses and unites the damaged portions. In wild animals where no setting of the bone is attempted this frequently results in a grotesque malformation. In animals where some setting of the bone is possible the repair is much neater.

The changes in internal architecture related to the bone as an organ do not concern the external morphology. The role of the bone in mineral metabolism, the effects of diet, pregnancy and lactation all influence the living bone and a great deal might be learnt from a study of these events on the architecture of the bone. The ability to recognise these events in archaeological bone would be of immense value.

Disease and infection can also alter both the internal and external form of the bone. Such pathological changes frequently involve both the erosion and deposition of bone, the latter being produced by the affected periosteum. Where this involves joint surfaces fusion (ankylosis) may occur. These disease processes are discussed in the

Figure 5 Epiphyseal fusion in limb bones.

(a) A femur in which both proximal and distal epiphyses are fused; in (b) both are unfused. (c) A humerus in which the proximal epiphysis is unfused. In animal (d) the epiphyses are fused. (e) A tibia with both epiphyses unfused. In (f) both have fused but the junction of the proximal epiphysis with the shaft is still apparent. In each case the deep aspect of the epiphysis is shown.

chapter on pathology. It is, however, important to bear in mind the three possible reasons for the deposition of extra bone—namely, normal remodelling, mechanical damage, and disease or infection. One or more of these factors may be involved at the same time and site and thus diagnosis is best left to a pathologist.

Teeth

The anatomy and functional characteristics of teeth are too complex to be dealt with in the present volume. Teeth are, however, of particular importance on archaeological sites for the information that they can give about the number of animals present and their age. Teeth are one of the most resistant parts of the skeleton to post mortem decay, and in certain soils their significance is correspondingly increased. The anatomy, development and functional characteristics of human and animal teeth and associated structures are dealt with by Scott and Symons (1958) to which the reader is referred.

Physical and Chemical Properties of Bone

A fresh bone consists of both organic and inorganic material. The organic constituents supply the physiological needs of the bone and provide toughness and elasticity in the structure whilst the mineral provides hardness. If a fresh bone is burnt to remove the organic matter it does not alter its shape apart from any distortion due to temperature. It does, however, lose weight and becomes very friable. Conversely the removal of the mineral by decalcification leaves a bone of normal shape but with the consistency of rubber. A classic demonstration of this property is to tie a treated long bone into a knot.

The chemical composition of bone can be determined in several ways according to the aspect of composition or structure with which one is concerned. Thus the chemical composition of fresh bone is different from that of macerated bone. Analyses will also differ according to the amount of each structural form of bone present in the sample. Thus in fresh bone the chemical composition of a whole bone varies according to the bone used, the species, age, sex, physiological status and health of the individual. Percentage compositions will also vary according to whether weight or volume is used and the chemical form in which the substances were determined. It is not surprising that the literature on the chemical composition of bone is confusing.

Archaeologically our concern with the chemical composition of bone is in relation to the preservation of bones in the soil. We are therefore not concerned with analysis of the total living bone, but of the structural elements. Fresh bone has three components: inorganic mineral which is physiologically active; inorganic mineral which is not normally active in ordinary homeostasis; and an organic fraction. The organic fraction comprises cell tissue concerned with the maintenance of physiological processes and collagen, a fibrous protein which is concerned with the structural arrangement of the bone. The collagen is resistent to decay after death and may survive for thousands of years (Garlick, 1969). The remaining organic matter undergoes autolysis after death and is rapidly decomposed.

The structural characteristics of individual bones are important in that they affect the permeability of the bone. The chemical composition of the structure is important in relation to the chemical reactions that can occur with percolating substances. Also important are trace substances and insoluble organic degradation products which may catalyse or inhibit further chemical reaction.

The structural arrangement of bone mineral consists of minute crystals arranged about the bundles of collagen fibres. These crystals are now generally considered to be hydroxyapatite, $Ca_{10}OH(PO_4)_6$. The surface area of these crystals is large and it is probable that other minerals present in bone, especially calcium carbonate, may be bound to the surface of the crystals or occur as separate amorphous phases. At the molecular level the structural relationships between the major mineral—hydroxyapatite—and other minerals present either in amorphous or liquid phase is not established in detail. Discussion of some of these aspects will be found in Wolstenholme and O'Connor (1956); Rodahl, Nicholson and Brown (1960) and Evans (1966).

Factors Influencing the Survival of Buried Bone

Before a bone is discarded several things may happen which will affect its ability to survive burial. When cutting a carcass for meat the bones of different species are treated differently. Sheep and pig joints are generally cut and sold with the bones, whereas beef is more frequently sold filleted. This practice reflects the size of present-day households and the price of meat. Most families require only a relatively small joint so that the butcher cuts a carcass with this in mind. The frame of a sheep is quite small and a major segment of the carcass such as a shoulder or half leg is an adequate joint for most families. Similar factors no doubt influenced butchery practices in the past. Although the general principle of small family, small joint holds true, butchery practices vary in different parts of the country and also between countries.

The bones that are sold as an integral part of the joint such as a shoulder of lamb suffer the same treatment as the meat. The joint may be roasted, stewed or boiled and the bones are then used for soup or discarded. The bones that are filleted out by the butcher are normally sold for further processing such as the extraction of fat or gelatin. Being a valuable commodity they are generally boiled or rotted before being discarded by the processor. The disposal of the waste bone will vary. At a household level many would no doubt be given to the family dog whilst others would be thrown on to the floor, fire, midden or into a refuse pit. From here they may be removed or partly eaten by scavengers. On a large scale the processor would probably dispose of the residue by tipping it into the nearest hole or on to a heap.

The chances of survival of the bone after disposal will depend on the physical and chemical properties of the bone when discarded and the nature of the new environment. A bone that has been roasted within the joint may have lost much of its organic matter and generally speaking is very brittle. The same is partly true for a bone that has been stewed or boiled and strength will depend on the type of bone and the length of cooking. Cooking for just long enough to make the meat tender does not render the bone very brittle though it will have lost some organic matter especially fat. If, however, boiling is prolonged it is possible to destroy the bone almost completely. A bone rescued from the stewpot at the last minute is little more than a brittle skeleton of its former self. The time necessary to produce this effect does of course vary between bones and between animals. It is the tabular bones and tabular processes that are the most susceptible. There is, however, little to be gained in the culinary sense from cooking bones to destruction. Thus much of the bone discarded from a kitchen is in the lightly casseroled rather than the heavily boiled category.

Commercially discarded bone will generally come from a boiling or rotting process. It should however be noted that from the early nineteenth century bone was being used as a fertiliser. It might be thought that the boiled refuse would be near the point of destruction. This is unlikely as the purpose of boiling is to extract the fat and gelatin. A high recovery can be obtained without prolonged cooking and it is probable that the contents of the vats rarely went beyond the point of moderate boiling. In these processes we are dealing with much larger and stronger bones since the majority will consist of the long bones of cattle. Waste derived from a rotting process is largely confined to horn cores of cattle. Parts of the skull might be left to rot so that the outer keratinous horn can be easily stripped from the bony core. Once the horn is removed the bony core is waste. The remaining occasions are those when animals are buried whole or a dead animal such as a dog or cat is left to decompose.

There are five main conditions in which bone can be discarded; fresh (probably rare in advanced societies), putrescent, roasted, lightly boiled, and heavily boiled. The first two are highly organic and the bone retains its physical properties. The roasted bone has lost much of its organic matter and structure and is in consequence brittle. The lightly boiled bone has also lost some organic matter but is still greasy and less brittle than the roasted bone. The heavily boiled bone has lost the greater part of the organic constituents and is exceedingly crumbly, brittle and porous. Given the condition of the bone when discarded we can to some extent predict what will happen to it when buried in the soil. Excluding now all effects of scavenging and gnawing we shall assume that our bone has ended up on the ground. If it went into a refuse pit or on to a midden it will probably be buried quite rapidly. On open ground it may lie there for an uncertain length of time depending on the amount of traffic, the weather and surrounding vegetation. The mud bath of many rural settlements in winter results in much churning of the soil and rapid burial. In a wood carcasses are very soon buried by falling vegetation and incorporated in the ground litter. Under any of these conditions some bacterial decomposition of the organic matter will occur and there may well be some effect from organic acids derived from vegetation and damage from the weather especially by frost. The most damaging action is, however, physical and will affect the more brittle bones which may be crushed under foot.

Once incorporated in the soil further degradation will be a direct result of the character of the soil. Thus the acidity or alkalinity, degree of aeration, movement of water, bacterial population as well as the structure and seasonal properties of the soil will come into play.

Most soils are aerated and here bacterial action is fairly rapid and will destroy much of the organic matter. Some aerated soils such as acidic sands may be less biologically active and organic decay (but not mineral solution) will take proportionately longer. In some waterlogged soils, especially those of very fine texture, bacterial decomposition may be almost completely arrested and decay will proceed over a very long time. The most commonly encountered examples of this are waterlogged cesspits and well deposits. Bones from such deposits often have a very fresh look about them.

Under very exceptional circumstances the bone may be preserved with much of its organic matter intact. Examples of this are the deep-frozen mammoths from Siberia, oilfield seepages, high aridity resulting in desiccation, mummification and the preservation of corpses in peat bogs.

It should be borne in mind by the excavator that the mere fact of discovery of material preserved under these exceptional conditions will set in play a series of new degenerative changes. Therefore steps must be taken at once to arrest these and to preserve the specimen as nearly as possible in the condition in which it was found. One possibility would be to deep freeze the material since preservatives such as formaldehyde might well render biological tests impracticable. In these circumstances it is necessary to get the best possible advice at once.

Generally speaking we can regard the stage of degradation of the organic matter as occurring fairly rapidly after burial. The most significant changes will be those that occur later associated with the dissolution of the mineral constituents.

Acidic soils will tend to dissolve away the mineral and the rate at which solution proceeds will depend on the degree of acidity and the amount of percolating water. Thus solution is most rapid where the soil is composed of non calcareous sands and gravels or is derived from acid igneous rocks, often in areas of high rainfall. I have excavated a number of sites on the gravel terraces of south west Essex where the rainfall is now about 30 inches per annum. These sites ranged from Roman to medieval times and I found that the bulk of the bone had been destroyed. Only the most resistant parts of the teeth and the odd pieces of bovine metapodials and other dense limb bones had been preserved. Thus it is clear that with this differential preservation of the evidence study of the bone must proceed with caution if misleading results are not to be obtained. A site in the same area at Mucking, excavated for the Ministry of Public Buildings and Works by Mrs M. U. Jones, showed marked differences in the preservation of bone on different parts of the site. This was associated with localised soil conditions. Whilst most bone in the sand and gravel had been destroyed that lying within pockets of fine loam was much better preserved.

Similar destruction of bone occurs in acid clays though here the rate of destruction is generally slower owing to a different water regime. On data such as those above I have come to the general conclusion that on acid soils it is unlikely that bone will survive more than a few millennia and in some areas for only a few centuries. Despite these observations it will be apparent that bone sometimes has survived in good condition in sands and gravels over a much longer period, the best examples being the mammalian bones found in river terrace deposits. In the case of the Thames, the upper catchment of which is in Chalk or Oolite, the gravels (as at Swanscombe) are often calcareous. Its Wealdon tributaries (Wey, Mole) often have no bone preserved.

Soils which are base-rich are much more conducive to the preservation of bone since they lack the dissolving acid solutions. Chemical changes do occur to bones in these soils and minerals are undoubtedly lost but the processes are relatively mild. One of the active agents attacking the bone is the roots of plants and it is not uncommon to find the roots of a plant wrapped round a bone fragment. This is also the origin of the etching and discoloration found on some bones which bear a superficial resemblance to blood vessels.

Despite these generalisations it is not possible to be too sweeping about the effects of a given soil/water regime because the factors involved are numerous and complex and may have changed with time. In general we should bear in mind that all soil conditions tend towards physical and more particularly chemical modification. These changes can be constructive as with petrification or destructive as with leaching by percolating acid solutions. It will be appreciated that when discarded those bones having the greatest surface area and porosity will suffer more than those which are relatively impervious. The bones that were friable and porous as a result of heavy cooking will be the first to disappear whilst the more compact and the least cooked ones will take longer. Again we have the factor of differential preservation associated, with bone type, species, and the amount and character of cooking. It is therefore important to examine the bones when found to see if there is any evidence of differential decay. This should be detectable as a result of the different physical condition of some fragments and after a long time in the ground by the discrepancy in numbers of highly susceptible and less susceptible bone types. In using this latter technique one should, however, take into account the other factors discussed in the chapters below. On none of the sites that I have examined where the bone is well preserved have I noted anything which has suggested differential preservation due to possible cooking effects.

Given the wide range of deposits that may be encountered there will occasionally be oddities of preservation that would repay fuller investigation. One puzzling example is occasionally seen in Roman and medieval bones from waterlogged deposits in the City of London. Some of the long bones when found are much blackened and appear strong and robust. If, however, they are allowed to dry out they become brown and begin to distort tending to split along their length and giving the appearance of being constructed from longitudinal fibres like a bamboo cane. The type of deposit is a waterlogged black sediment which also preserves leather and wood. Such waterlogged deposits in which leather is preserved are often strongly acid (pH < 5·6). In such cases collagen in the bone is preserved but the minerals will dissolve. The shrinkage on drying is due to that of the organic fraction where the mineral framework has been degraded. Bones that have been waterlogged often have the appearance of being in excellent condition when first found. This is almost invariably deceptive as they usually distort and split as they dry out.

The other effects of burial on bone are physical. In clay soils which dry out seasonally fissuring occurs and this invariably breaks up even the strongest bones. I have seen long bones of cattle from acid clays where the bone is friable but has preserved its form. The expansion and contraction of the clay has however broken the bone into small fragments. The same damage can occur when bones, especially skulls, are lifted and left to dry out full of soil. Frost fracture and frost heaving both in antiquity and at the present day also shatter bones. Soil creep and solifluction will also move and damage bones. Where the soil is relatively stable the commonest cause of damage is crushing from the amount of overburden. Pressure and slight lateral movement combine to crush and shear a bone resulting in the lateral displacement of fragments. These forces can also cause distortion of the bone and great care is necessary in interpreting cranial reconstructions which are made from part of the cranium. When reconstructing some of the Saxon dogs from Mucking we were unable to unite adjacent bones in the anatomical position because of a shearing distortion of the bones. Had we had only a portion of the skull we would have been able to reconstruct the skull quite easily. Unfortunately this reconstruction would have been incorrect as we would have reproduced the distortion and never have known otherwise. Having the whole skull in fragments enabled us to detect the distortion.

Despite all these hazards an enormous quantity of bone survives for us to study. There is, however, much to be learnt about the processes that go on in archaeological strata and which ultimately leave us with our sample. I therefore urge both archaeologists and specialists to look a little more closely at the conditions under which bones are or are not preserved; the findings may be unexpectedly rewarding.

2
Planning and Organisation of Bone Studies

Bone Studies in Archaeology

The value of the biological material from excavations and some of the problems of dealing with it have attracted attention for many decades. Pitt Rivers was one of the earliest excavators to appreciate the potential of animal remains and to begin to study them. His work was not surpassed for many years. The papers by Hargrave (1938), Gilmore (1946, 1949), White (1956), Reed (1963), Boessneck (1964) and Chaplin (1965), reflect the views of biologists, archaeologists, anatomists and anthropologists on the value of this material.

Bone studies are no different from other specialised studies which are an established part of the archaeological routine. The study of the bones can only be carried out by someone trained in the appropriate techniques and not by someone armed with a textbook and a rudimentary knowledge of anatomy. All too often a specialist is called in only when the excavation is finished so that much of the material is not suitable for detailed study. As each specialist will have his own requirements for the collecting and recording of the bones it is essential that the bone study should form a part of the initial planning of the excavation. Some general considerations will, however, act as a guide to the way in which the organisation of bone studies should be approached. Excavations abroad frequently present additional problems to those at home and these particular problems are discussed below.

The preliminary planning of the excavation should cover the following points:

1. The provision of a bone specialist and any ancillary bone staff.
2. The degree of on-site work to be done by the specialist or his delegate.
3. The field routines and requirements for recording, collecting, preserving and transporting bones.
4. The finance of the field and laboratory work.

In seeking a bone specialist and finance for his work it is important to remember that very few specialists are employed as such. The majority of specialists are members of university and museum staffs or amateurs and each has his particular problems when undertaking site studies. Some specialists are interested in studying a site collection for information about the animals, economic activity, and ecology of that site, while others are interested in only a part of the material for a particular research project. It is very desirable that the former should examine the collection first and then pass on appropriate material to the latter. Where the reverse happens and the site material is divided up it may be difficult to re-assemble it for a full site study. My own requirements for the collecting of bones and their subsequent study have been standardised and the same system is followed by several other workers. These requirements are set out below; they are fundamental to almost any bone studies and have generally been found satisfactory. I would emphasise that any lesser standard seriously detracts from the ability of anyone to draw useful and valid conclusions from the bones. Undoubtedly these standards may be redefined in the light of fresh thinking and others may have more specific requirements. It is therefore essential to find out what the specialist requires and only to fall back on these minimal standards in an emergency. The desirability and practicability of on-site work by the specialist or an associate are rarely the same although it is important that the specialist should spend some time on the site if at all possible.

An important consideration is finance. Virtually all bone studies in Britain are heavily subsidised by bodies who have little direct concern with the site in question. One hesitates to cost the matter in public but it is important to realise that this subsidy does exist and is considerable. That it exists is a fine example of the liberal approach of so many authorities to work that is basic to an understanding of our cultural and economic history. It must be remembered that the bulk of this subsidy comes in the form of staff, facilities and collections deployed over a much wider field. In consequence the more use that is made of these facilities the better. It can, however, be rather a different matter to find cash from official funds to finance visits to a site or to purchase special materials. All too often the burden falls on the individual concerned and one knows that few can afford this. The problem is of course greater for the amateur worker who in theory must find the majority of his own expenses. In fact he is often paying for the privilege of doing the work. There is therefore an urgent need for an appropriate sum of money to be made available to the specialist. This must be sufficient as a bare minimum to cover visits to the site (travel and expenses), any materials supplied (boxes, bags, chemicals), postage, etc, and production of the report, including commissioned drawings. Out of fairness to all concerned every excavation grant application should make provision for the bone studies. I know that in the past with some funds to meet expenses attached to each site the productivity of my own students could probably have been trebled. As it is, one is prevented from making full use of the people that one has trained. In approaching a university department one is dealing with a totally different organisation from the semi formal group system which is developing in Britain in response to the need for many more workers to deal with the routine material. The university worker—and his ancillary aides, principally full-time students in training—must often be more concerned with particular types of sites and problems. He should not be expected to tackle routine work unless it bears on his particular field of interest. In regard to the finance of investigations his position can be rather different. If the site in question falls within his field of research it is likely to be supported by grants for that purpose. Further, in most universities there are sources of finance which may be available to meet specific requirements. Generally these are not accessible to the outside worker, who in order to obtain funds must often compete with more orthodox projects for the limited funds of a particular grant giving body. This is not to say that such bodies are unwilling to support the work of the bone specialist but rather that they are very often already strained. The question of the provision of such fundamental items as comparative collections and X-ray facilities is something that the specialists must face themselves. It need not concern the excavator.

The specialist abroad faces very different problems compared with those from sites in his own country. It will be apparent in due course just how dependent the specialist is on zoological research. Many sites abroad however are in countries where almost no work has been done on the native species and very little if any comparative material or data may be available. The position of the specialist in these circumstances is a difficult one. To do his job properly, not only will he need to perform his on-site duties, dealing with the bones found, but also to study in a short season the animal ecology of the area. It is this, I think, that adds force to Reed's view that 'osteo-archaeology' is a task for the zoologist. In these circumstances I would agree with Reed (1963) who has faced this problem in his work on domestication in South-West Asia. What really is the task of the specialist when faced with a site largely unknown zoologically? To deal intelligently with the excavated material he must obtain modern comparative animals from the area and appreciate the ecosystem and how and why it may have changed with time. This is not a task for one individual but rather one for a team, some dealing on site with the problems of excavated material, others in the field collecting, recording and carrying out the preparation of carcasses. It is not a task likely to be done properly in a single season nor done without careful planning and provision. Frankly one must ask whether the collecting of comparative material and the analysis of the environment should form part of the archaeological programme. It seems to me that the answer must lie in inter-disciplinary expeditions involving specialists in zoology. botany, geology, etc, as well as archaeology. In this way one is assured that all tasks will be done thoroughly. It has never fallen to me to have to attempt such a project. The zoological side of this is, however, well known to me in Britain where the handling of both large and small carcasses in the field and laboratory for intensive zoological study and skeletal preparation has become a routine. To deal adequately with this material some simple but highly efficient techniques were evolved which may be of help to others in the field. These are described in Chapter 3.

One of the difficulties of operating abroad is that the worker is often divorced from much of his ancillary equipment and sources of reference. This is no handicap if he is able to remove the collection for study. This is not always so and often the work must be done in the field. It is an unfortunate state of affairs when this occurs for bone studies are not accomplished in the short season of a dig. Even if one is able to remove the collection it must often be drastically sorted because of the expense of transport. Done properly this has the disadvantage that the repair work may be hampered but it will not distort the results. On the other hand inexpert sorting will introduce an unknown degree of selection bias which renders material useless for many studies. Yet if, as often seems to happen, there is not a specialist available on the spot it almost seems preferable that the bones should be reburied intact in stratigraphical groups rather than that a lot of nonsense should appear in print increasing the myths which are already widely disseminated through the literature. Bones buried in this way can be recovered at an appropriate time. In a very difficult situation it may be permissible to jettison some of the bones. This may be done with strata where the degree of contamination from earlier horizons as indicated by the pottery, etc, is high. However, anything of zoological interest should be brought back but the remainder buried otherwise intact for later recovery and study. One should not ignore the recent horizons for the changes in the economy and the livestock with time are of particular interest especially when they can be brought up to the present.

Requirements for Collecting Bones in the Field

The investigator must satisfy himself that all his conditions are being observed in the field and are not being sacrificed to expediency. The following notes are concerned with the minimum standards established for my own purposes and which are common to a number of colleagues. I hope these will enable the excavator to assess his task and also demonstrate the reasons behind these requirements. As an excavator myself I am able to appreciate the problems of the excavator in the field so that these requirements are as practicable as the subject permits.

Every scrap of bone found must be kept for examination. There is no room for argument on this point. Unless this is done no one will be able to produce valid information.

The best method of collecting and handling bone on the site is to equate bone with pottery and to handle it in the same manner, that from each particular pit, level or feature being kept separate. Subsequent study and discussion with the specialist may indicate that some groups can be amalgamated. It is normally the pottery and other artefacts which date and may define phases of the site:

in the final analysis it is to these phases and dates that the bone material will refer. It is important to establish for each group or level the degree of contamination from material surviving from earlier levels; artefacts provide a means of doing this. It is clearly false to determine an economic pattern for a phase of a site when the bone collection on which this is based contains a large number of bones from an earlier and distinct phase. Such contaminated groups are best accumulated but should not be discarded without examination in the field in case they contain zoologically important specimens or other fragments which might help to reconstruct an important discovery from an earlier level. However, the conclusions that can be drawn from such specimens are limited by the degree and nature of the contamination of the level in which they were found and the significance must accordingly be established with this in mind. It follows from this that closely dated groups of uncontaminated material have a special significance and where large quantities of bones from such deposits are available their excavation must be accorded the highest priority and it may be worthwhile to extend trenches in order to obtain the largest possible number of bones from that horizon. The importance of large groups for statistical purposes cannot be over-emphasised.

The analogy with pottery and other objects may be taken further to demonstrate the relative importance of different pieces of bone and the way which they should be handled and recorded. The usual broken pieces should be collected and handled as sherds of pottery would be. The complete bone should be treated as a small find and be marked and recorded accordingly though if it is safe to do so it may be kept with the smaller fragments. This provides a certain check on its provenance. The same technique of recording applies to articulated bones, skulls and whole skeletons. These are of extreme importance for so few are available for study. The technique for dealing with these is given below but they should normally be lifted undisturbed in their block of soil and sent to the specialist for dissection.

The presence of very small bones, of such animals as mice, frogs, fish, etc, is often overlooked with normal methods of digging and it is necessary to carry out sample sieving as the deposit is worked through if these are to be recovered. The quantity sieved should bear a known relationship to the amount excavated and the samples should be taken so as to ensure that the whole level is sampled both horizontally and vertically. The proportion sampled must be decided in the light of practical problems for it is desirable that all should be sieved in this fashion. Although these techniques are more frequently practised on prehistoric sites it must be emphasised that they are equally applicable to, and the results of interest on, later sites. Site sampling has been carried out as an excavation method on prehistoric sites in America and is described by Treganza and Cook (1948) and Heizer (1960).

These techniques, however, are unlikely to become commonplace on many sites. Although to produce quantitative results comparable with the other bone material, proper sampling is necessary, from the zoologists point of view qualitative records are of interest. Rather than that these records should be lost, the sieving of about 1 cwt of soil from each horizon is recommended as a routine practice on all sites.

The material that will be recovered from sieving is not of course confined to bones. Seeds, wood, molluscs, remains of beetles, etc, will all be recovered if soil conditions are conducive to their preservation.

Another aspect which has been little studied is the relative volume of bones to other refuse in different deposits. This is an interesting line of enquiry which deserves further investigation.

Articulated bones, skulls and whole skeletons require particular attention. Bones found in an articulated state indicate that they were discarded or buried with connective tissue or flesh still in place. The information which can be obtained from bones known certainly to come from the same animal is far greater than from large numbers of fragments. These bones can be either lifted in place on a block of soil or packed together. Either a sketch or a photograph (within the limits noted below) showing the relationship to other finds should be made. With skulls and whole skeletons only the minimum of soil necessary to delimit the body should be removed in the field. The skeleton should be lifted on a rigid base and left for subsequent examination in the laboratory where it will literally be dissected out by the specialist. The reason for this is to avoid damage to the parts and to recover other material from the adherent soil which can only be done under laboratory conditions.

The greatest enemies the specialist has to face are cleaning for photography and faulty impregnation. First of all we must consider the damage that bones suffer through being cleaned for photography in the field. Individual bones vary in their susceptibility but all suffer a greater or lesser amount of damage. The purpose of cleaning for photography is to reveal the object with the utmost clarity possible. It involves a lengthy dissection and brushing which in the field will often cause mechanical damage. It is further a lengthy task and inevitably delays the excavation. Far worse, however, is the fact that moist bone is being exposed to rapid drying through the action of wind and sun. This occurs only on the exposed face of the bone, the other, being in contact with the soil, remains moist and stresses are set up which lead to the splitting or warping of the bone. Whilst this can be overcome by the application of water it seems a rather lengthy process just for the sake of a photograph of a skeleton *in situ*, under difficult conditions. A skeleton can be lifted very rapidly *en bloc* without delaying excavation work. If a photograph is desired it can be produced under ideal conditions in a laboratory without loss of other important information. A whole burial may be expected to yield to detailed scrutiny in the laboratory, traces of hair, external parasites and flies from the soil immediately over the body. From inside the body may come botanical material from the animal's food, e.g. seeds, and also the eggs and body parts of endoparasites. All this material is liable to be lost or contaminated if the skeleton is cleaned in the field.

Any skeleton or skull lifted with the soil in place, whether it is already broken or not, needs very careful attention. The general treatment given to these bones in the field is to apply polyvinyl acetate to the surface to bind the fragments together and solidify the whole mass. Bone so treated is of course useless for Carbon 14 dating. It has been my unfortunate experience that where this treatment has been given by both trained and untrained people, myself included, it is virtually useless and induces a sense of security which is often rudely shattered. The only value that I can see of the application of plastic compounds is to bind in the surface fragments. As soon as a soil dries out it contracts, a pliable soil in a strong skull will go to powder and do little harm to the bone, at the other extreme a tenacious clay, on drying, will pull a skull to pieces. The rule with these bones is that once lifted they should be kept wet and taken to the laboratory as soon as possible where the necessary work can be carried out. No bones should be dried rapidly by placing over convector heaters, in direct sun or in a warm dry wind. Rather they need to be left to dry out slowly to prevent distortion.

Distorted bones are often useless. Impregnation with plastic whilst bones are still wet can only be done satisfactorily under vacuum conditions. Normal plastic solutions will not penetrate wet bone and an emulsion must be used.

Some workers require the bones to be washed clean, others prefer them to be dry brushed. Both of these requirements are valid. The purpose of having the bones unwashed is to avoid contamination where biochemical examinations may be required. Normally this can only be carried out satisfactorily on intact bones. For my own purposes I require all the bones to be washed free of adherent soil. Washing does of course take time but I am not prepared to spend my own time washing large quantities of bone fragments. Since much of the fragmentary material is not the subject of detailed scrutiny there is no need to have all these fragments spotlessly clean. All that is required is that most of the dirt should be removed.

For many purposes bones that are strong and well preserved can be dealt with in bulk rather than individually. Material that will not stand handling and washing should be dealt with individually and details of treatment which can be applied are given below.

A quantity of bone requiring washing may be dealt with as follows:

It should be loosely packed in a suitably large container such as a wheelbarrow and the end of a hose placed in the bottom. Water is then allowed to fill and overflow from the container and washing is continued until the water comes out clean. Gentle agitation will help to free lumps of dirt. Most material washed in this way will only require a slight brushing to remove the greater part of the dirt.

Fragile bones which do not require chemical treatment should be washed individually with a soft brush in water.

The addition of detergent materials to the washing water should be avoided as this will render the bones unsuitable for a number of physical and chemical investigations, e.g. blood grouping (Brothwell, 1965).

Some bones of course will not withstand washing; where the strength of the bone is in doubt a small and unimportant fragment should be tested, and each fragment washed individually. Where bone is in a very friable condition and requires a great deal of treatment to preserve it one must question whether this elaborate treatment is justified by the results likely to be obtained. Very often with sites of a commonplace character the answer must be "no" and it is here that the bones can be ignored. This should not be an arbitrary decision by the excavator but be made in consultation with the specialist. Although it may be impracticable to retrieve the material, much can often be identified *in situ* as the excavation proceeds.

Whole Skeletons

The discovery of a complete or semi-complete carcass is of extreme importance to the specialist and it is desirable that as little as possible should be done to this in the field. There are many reasons for this and if the fullest possible information is to be obtained from the skeleton it is essential that the following procedure should be carefully observed. When it is thought that a possibly intact skeleton has been located the further clearance and lifting should be handed over to someone who is thoroughly trained in the techniques of excavation. The object of the exercise is to lift the skeleton intact with the removal of the absolute minimum amount of soil covering it. It will obviously be necessary to locate the extremities of the skeleton so that the area in which it lies can be determined. The head at one end and the tail at the other will normally define the overall length of the carcass. The spines of the vertebrae will generally define its extent in one direction and the toes or lower limb bones in the other. These must be located not by continual clearance and tracing of exposed bones from one end of the body to the other but by the inspection of very small test holes or by gentle probing. The ribs and pelvic area in particular should not be exposed. The reason for this is that much of interest may be contained in the soil within a few inches of the body. If a photograph of the skeleton is desired this should be stated since in the laboratory the specialist can dissect out the animal and then take the necessary photographs. An orientation should of course be marked on the block of earth as well as a basic datum point. In general, however, the specialist will not need to be asked to take photographs of the animal as it is dissected out as this will be routine in most laboratories.

The lifting of such a skeleton is quite straightforward. Having defined the limits of the skeleton and the position of the principal limb bones the soil should be removed up to six inches from the edge of the bone and cut down from a foot to eighteen inches depending on the nature of the skeleton present. The skeleton is lifted by driving a sheet of steel with a sharpened edge through the soil supporting the bones. Having truncated the supporting column the steel plate is lifted with the skeleton on it. If necessary this should be slid on to or replaced by a solid support of timber to avoid buckling and sagging. The skeleton should now be transferred to the laboratory as rapidly as possible. If this is a matter of only a few hours no further treatment will be necessary. If, however, the laboratory is reasonably close but some delay is expected it is necessary that the soil and carcass should be kept moist and this is best done by placing damp newspaper or similar material over the surface. It is a sensible precaution to do this anyway as it will provide protection to the few pieces of bone which are actually exposed. If, however, the base laboratory is some considerable distance away and it is necessary for the skeleton to be shifted by contractors it is generally desirable that the steel plate should be removed so that it can be used again and a crate built around the block of earth, the top being suitably protected. Particular care should be taken that the bottom of the box is reinforced and that it will not fall away as the water drains out from the top.

The task in the laboratory consists basically of dissecting out the skeleton from the soil and the continual testing of the soil for other biological specimens in particular the remains of beetles, fleas, ticks, seeds, plant debris, etc, and their precise recording in relation to the different parts of the body. As already mentioned photographs will need to be taken at various stages. It goes without saying of course that objects of archaeological interest will be recorded in similar detail. The detailed examination of carcasses such as these is as yet by no means routine and indeed many of the methods for recovering other zoological and biological material have not been worked out in detail, so the specialist will often find himself carrying out investigations on how to recover this. A great deal of assistance, however, is likely to be obtained by consultation with parasitologists who will be familiar with the fauna that may be found in or on the animal concerned and the type and size of specimen that may be recoverable. Although one would not expect to recover tapeworms intact except in the most exceptional circumstances it is by no means impossible to recover the eggs of these and other endoparasites. Similarly the large swollen bodies of ticks will not be recovered but the minute mouth parts and legs may well be preserved in the same way as the elytra of beetles. It is up to the specialist to decide just how far he carries investigations such as these but it seems to be one of the most promising fields of investigation. At the present moment it may well be the only means we have of estimating the health and condition of many of these animals and the type of habitat in which they had been grazing or living.

Labelling, Packaging and Transporting Bones

A great deal of post-excavation fracture occurs when the bones have not been carefully packed for transit. This alone is sufficient to invalidate the method of counting the number of fragments which is used by some workers to obtain the ratio of species. Other objections to this method are given below. Also, if careful attention is not paid to labelling and the type of bags or containers used, much material may lose its provenance through the labels becoming separated, decayed or the containers ruptured. A great deal of time and work can be saved if careful preparation is made in the field.

Fragile bones from any part of the body will need particular attention. Assuming that they have been consolidated in the field each should be padded and packed tightly in a rigid container. Individual labels are attached to the specimen beneath the padding. The padding is held firmly in place by string (not rubber bands or sellotape which perish) and a duplicate label is threaded on to this so that the specimen's provenance and nature is identifiable without disturbing the packing.

Bones which by their size or nature are delicate, such as large and small skulls, fish bones, etc, require individual attention. Small bird or fish bones, small mammal bones, etc, should be wrapped in paper (paper handkerchiefs are ideal) and placed in rigid containers such as tins, polythene tubes, stout cardboard boxes, etc. These bones if left free to roll around in the containers will be damaged. Skulls, etc, need packing both to prevent movement in the containers and to give support to avoid undue stress. Glass containers should not be used; the glass tubes that are sometimes used for small bones are very easily shattered. For large skulls a shock-absorbing, but semi-rigid bed to provide support should first be prepared on the floor of the box. The skull is then put in place and a filling material such as wood straw, newspaper, etc, is placed firmly around it. Sawdust and insulating materials such as "Micafill" have their uses but great care must be taken to ensure that settlement will not occur. If available it is a good idea to line cardboard boxes with expanded polystyrene foam which, whilst giving protection, is sufficiently soft not to damage the specimens if movement does occur. As much damage can be caused by movement within the container as from blows from outside, so a foam polystyrene lining is a good insurance against both.

Intact long bones should be wrapped individually to avoid abrasion and packed firmly into suitable stout containers. The remainder of the bones which will comprise the bulk of the collection may be treated slightly differently if they are in good condition. Either boxes alone or bags in boxes may be used depending on the size of the group to be kept separate. The greatest enemy of packaged bones is weight, either from weak boxes stacked one on top of the other or too much bone in one box. A tea chest full of bones weighs a great deal and the bottom layer of bones receives most of this. Packed in this way it is not surprising that the bones become more fragmented the deeper one goes into the chest.

One cannot specify an optimum size of container for bulk packaging but I prefer one $16 \times 10 \times 8$ inches. A standard size box enables the bones from a site to be put away immediately on to standard shelving. Non-standard boxes are a nuisance as they take up a great deal of space and the contents have to be sorted and reboxed before they can be put away. The standard boxes may be used to contain bags of bones. The fragmentary bones are placed in a large polythene bag inside the box as a safety precaution. Since none of the fragments has been marked about six should be marked with the full reference for the group. A plastic label of the type used in gardening bearing the same reference should be placed *in* the bag. Having settled the polythene bag in the box the neck of the bag is closed with string and an identical label tied on the string. The box is marked on one end with details of the contents and the lid is sealed down with sticky tape. Ideally the list of contents of any box should be enclosed in a polythene bag.

These requirements have been worked out as the result of bitter experience from many sites. Everyone believes that his packaging, etc, is adequate—but all too often it proves not to be so. These requirements are an insurance against the deterioration of packaging materials, clumsy handling, and the inquisitive fool. All of these are familiar to museum curators. These problems are not peculiar to the bones and a more careful examination of the reasons for these precautions may be of value.

Deterioration of materials is not a new problem but has become more acute as modern technology often reduces the quality of material in order to cheapen it. After all, who, except an archaeologist, wishes to keep a paper bag for twenty years? In a store which is reasonably warm and dry, the life of a stout but cheap paper bag is at the most five years. Bags of pottery filled longer ago than this cannot be picked up without their falling apart. Under damp conditions these bags are useless. Damp, either from the contents or from outside, reduces them to pulp. The same is true of paper or card labels. The bags used by the Ministry of Public Building and Works have a longer life but as these are supplied in a large size they are invariably filled with bones which are too heavy for them. With any paper bag a sharp pointed bone (and most fragments are) is sufficient to cause a bag to split. The result of this that one may be faced with a mixture of bones caused by the bags bursting. Double bags are a useful precaution but are not a satisfactory answer to all the conditions which may be experienced. The only satisfactory bag is of about 150–200 gauge polythene which is proof against almost all forms of decay or accidental damage and has a sufficient strength under all conditions to contain the material. The same deficiencies that affect bags are also found in many paper labels. Paper labels can often be improved by subsequent impregnation with a plastic solution. More satisfactory, however, is the use of plastic labels or corrosion-resistant metal tags such as are sold for garden use. The normal type of plastic label is not, however, satisfactory for use since it has a smooth surface from which the pencil or indian ink is removed at a touch; furthermore, the solvents which are used for polyvinyl acetate, etc, will dissolve the label, so preventing the use of a plastic seal to protect the writing. The ideal label is made from celluloid and has a matt surface on both sides. It is purchased from Hospital and Laboratory Supplies Limited, 12 Charterhouse Square, London, E.C.1, in sheet form and provides the cheapest means of producing high quality labels in quantity. It may be punched with a paper punch to provide tie-on labels and is virtually indestructible in the circumstances in which it will be used. It is normally written on in pencil, which cannot be completely erased owing to the matt surface, except by scratching away the plastic.

Rubber bands perish and no labels should be affixed with these. String should be used and this has a much longer life but will perish eventually. Generally string will be satisfactory but nylon lasts far longer.

Cardboard containers come in all shapes and sizes and whilst it is possible to obtain really strong containers made of card or paper they can be very expensive. The cheaper types of boxes normally used on excavations and by museums are especially susceptible to damp, deterioration and mechanical damage. Mechanical damage caused by overloading is of course avoidable; accidental damage is not, nor, often, is damp or deterioration. The collapse of a box for any of these reasons leads to spillage of the contents. This is normally avoided by having a polythene bag inside which carries the contents. It also means that the contents are visible for examination by all, from Customs officers to the excavator, without the risk of mixing material.

Unfortunately one cannot entirely remove the risk of mixing by some misguided person. Double packaging, the tying up of bags and the sealing of boxes will deter the majority of the curious. Further, these precautions guard against the very real problem of accidents, boxes falling off a lorry or being knocked over. When this occurs and specimens are spilled the natural reaction of the person responsible is often to put them back into any box and say nothing. Where bones are loose in boxes it may never be detected that this has occurred. Where bones are also packed in sealed bags they are much less likely to spill out and in any case a burst is readily detected and one is put on one's guard.

On other occasions each package may be opened and the contents examined. This is likely to occur when specimens are sent through Customs. It is imperative that where material is being sent across Customs barriers special arrangements should be made so that the material can pass through without examination or be examined with a responsible person present.

In short, scrupulous care in labelling, packaging and transit arrangements is essential if the specimens are to be in a fit state for any examination to be made.

Preliminary Tasks to be Carried out by the Specialist

Having considered above the work which will be carried out by the archaeologist in the field and in conjunction with the specialist we may now turn to the tasks which will be carried out by the specialist as an essential preliminary to the detailed study of the bones. Already during the course of the excavation the specialist, particularly where he has a laboratory available, will have received certain specimens which required immediate attention. There should not in general be a large number of specimens requiring routine vacuum impregnation in order to consolidate them. This is the responsibility of the archaeologist in the field or in his own laboratory. The specimens sent to the specialist for treatment are those which are of particular interest, or which will yield additional information if conservation is carried out by the expert. The sort of specimen that is best left to the specialist to deal with is something that is particularly friable or delicate and which it may not be possible to restore or reconstruct simply from a collection of fragments, e.g. skulls, all parts of which are present but badly broken, or, a large section of a long bone which is badly splintered but which, if carefully dissected out, could be strengthened and treated for subsequent measurement and study. The dissecting out and subsequent repair of these specimens is an extremely lengthy process and only if the importance of the material justifies it, should this be done. Decisions like this cannot always be made in the field and it is up to the specialist to examine the material and decide whether it is worth spending a lot of time and money in reconstructing a broken portion. In other cases where a bone or skull may be intact in the ground it is desirable to lift it in one piece and to take it back to the laboratory for examination. The behaviour of bones which have been in the ground for a long time cannot always be predicted. It is much better that these should be sent to the laboratory so that if it is necessary carefully controlled drying may be carried out to minimise any distortion. The value and treatment of entire carcasses has been described above. The conservation or restoration work that will be done by the specialist is to prepare the specimen for zoological study. It is another matter to conserve or restore a specimen for exhibition and although the two may not be incompatible it should be clearly stated whether a specimen is intended for subsequent display. The needs of the zoologist, however, should not be subordinated to the requirements of museum displays. One often finds that people who have expressed little interest in the large numbers of fragments which turn up on a site immediately seek to acquire any complete or semi-complete skull on the grounds that it should be on display in a museum.

It is likely that the remainder of the bones from the excavation will be received in one large single consignment. The way in which this is dealt with will depend both on the resources of the specialist and the amount of preliminary packaging which has taken place in the field. The specialist should be notified of the number and nature of the containers that will be arriving so that these can be checked off and any shortages notified. From here the material may be stored in its existing containers or transferred to a more standardised form such as has been described above. As far as possible associated material should be boxed together and a list of the contents of each box should be placed inside the lid where it is easily accessible. When transferred to final storage the contents should be marked on the outside of the box together with the name of the site. A record made at the time of receipt of the number of containers and the specimens therein can save a great deal of searching at a much later date. When received the specimens should be looked through briefly to see that they are in reasonable condition and that no important specimens have been overlooked which might require further treatment. This preliminary examination also gives a clear view of the nature of the material to be dealt with which will be of great assistance in planning the detailed examination. Having received the material the only other thing which remains to be done before the detailed studies can commence is to decide upon the programme of work and precisely what questions are to be answered and the methods that are to be employed. The specialist should have received a preliminary briefing on the deposits, their nature and significance, and also been given notice of the sort of questions that the archaeologist particularly wants answered. It is from these and his own thinking that the specialist will determine a programme of work which will be submitted to the archaeologist for his approval in detail.

3

Bone Identification and the Establishment of Reference Collections

Problems of Identification

Neither the recognition of particular bones nor their specific identification falls within the scope of the present work. Experience has shown, however, that a restatement of some of the problems associated with identification and the procedures that should be followed would not be amiss.

The identification of a fragment of bone proceeds in two stages. Firstly, the bone of which the fragment is a part must be identified. As a first step one may recognise it as a rib or vertebra but until its precise position in the body has been determined no identification as to the species from which it came can be made. Thus, whilst one may recognise quite readily pieces of the limb bones or girdles the rib or vertebral fragment must also be placed correctly. If a rib, one must recognise, for example, whether it is the 6th or 12th and if a vertebra, whether it be lumbar or cervical and as far as possible its approximate position in the lumbar or cervical series. Anyone who has sorted a collection of fragments will know that it is essential to have a complete skeleton to hand to assist with the task. Much can of course be accomplished if one has an excellent memory of anatomical features and it is true that in time one develops an instinctive feel for recalcitrant fragments. However, accuracy will be enhanced, and the pile of indeterminate fragments greatly reduced, if a skeleton is to hand on to which one can try and fit the fragments. Since the basic features of the mammalian skeleton are remarkably constant it matters little what animal is used for this purpose. It is, however, desirable that the prepared skeleton should be reasonably compact and the animal concerned should be easily obtained and prepared by the student. It will also be an advantage if the animal used is one that will be needed for the identification of the species found on excavations. The most suitable animal to my mind is a dog or a fox, there is little difference between the skeleton of the two and both are obtainable from veterinary or farming sources. When defleshed they may be cooked clean in a bucket or electric boiler. A hare or rabbit is also a useful animal but is rather small for comparative studies. In some respects a sheep is perhaps a more useful acquisition. It is fairly easy to recover a carcass from the hills where a great many die each winter and if one is fortunate the bones will already be partly prepared. Zoological gardens, parks and estates are also good sources of dead animals of all descriptions and the only limiting factor is the strength of one's stomach and the tolerance of one's colleagues or family. Bird bones to the uninitiated can cause confusion and the skeleton of a bird would be a valuable item. In my experience the best source of ligamentary bird skeletons is the upper debris line on a beach.

Having identified the bone one then proceeds to track down the type of animals from which it could have come. Having an idea of the size of the animal from which the unknown bone came a list of possible species may then be drawn up. The presence of distinctive features between the genera or families such as the development of the third trochanter in the horse may serve to further limit the possibilities. In the case of a femur showing a third trochanter from a Roman site in Britain cattle and red deer would have been eliminated on anatomical grounds leaving only the horse family. More often than not one is left with three or four possible species which must then be compared with the unknown bone and each other to find the diagnostic features. In many cases there will be clear cut differences and a knowledge of these will enable the specimen to be assigned immediately to a particular species. More frequently, however, the specimen lacks clear diagnostic features and it is only by matching against known bones that a determination may be possible. Often, however, the fragment may not have features that distinguish between several species and one ends up with a fragment which could be of horse, cattle or red deer. Such fragments should be classified as indeterminate and normally have no place in the calculations of the significance of different species. In certain circumstances it may be necessary to take these into account, so they will be recorded or set on one side. Only after the basic examinations have been made will it be clear whether these "groups" should be considered.

In comparing bones in order to make identifications several very important points need to be taken into account. If we consider any single species in one particular environment we are at once aware of the variation in that population both as to size and perhaps, especially with domestic stock, to colour and other pelage characteristics. These differences will not all be reflected in the skeleton unless there is some definite association between, say, fleece type and the anatomical features. The main difference that will concern us is size which is clearly reflected in the skeleton. This difference may be related to sex (sexual dimorphism) in species where the sexes are different. Superimposed on any sexual dimorphism will be the normal range of variation between the individuals in any population. It is to this normal range of variation and also to sexual dimorphism that particular attention must be paid. If we take a single bone of, say, an ox or a red deer we may find many slight points of difference between the two but it is only when these differences are borne out consistently by a much larger sample covering both sexes and different age groups can we claim to have found features which are diagnostic. The inviolable rule is that where differences appear to be but slight it is necessary to examine a long series of specimens to establish the constancy or otherwise of the differences noted.

In dealing with archaeological material we introduce a further set of complicating factors. Our archaeological specimens are samples of populations distant in time from those of today and possibly existing in very different environments. It is therefore not surprising to find very considerable differences in size and often in anatomical features as well between modern specialised livestock and the same species from an ancient site. Our picture is therefore of ancient animals often very different in size from the modern ones (especially with domesticated species), each having its normal range of variation and varying degrees of sexual dimorphism. Studies on the variations in a modern population under controlled conditions are fundamental to an understanding of the ancient collection and experiments with modern livestock must become an increasingly important feature of osteo-archaeology.

We do not yet know much about the variation (especially of the skeleton) of many modern species so the osteo-archaeologist cannot expect at this stage to be presented with a set of magic formulae which will perform wonders of separation. Many bones set severe limits to identification and these things must be accepted. Identifications must not be forced; if doubt exists it should be acknowledged or else removed by the study in depth of modern species. One must, however, be careful in extrapolating back the results obtained from modern animals. For example, with the sheep/goat question Boessneck, Müller and Teichert (1964) examined an extensive series of modern animals of different breeds, sex and ages and were able to point to differences between the species of varying degrees of reliability for almost all bones of the body. Several workers have attempted to apply these to ancient populations from excavations and have claimed that some of the criteria were not always recognisable and that only some of the features that were diagnostic in the modern sample held good for the ancient ones. Thus it must be accepted that many species will not be distinguishable from every bone and in economic analysis recourse will have to be made to those bones which are certainly diagnostic.

In making accurate identification there can never be any substitute for comparative specimens. Even the most comprehensive sets of illustrations will often not help in identifying the largely fragmentary specimens from excavations. Regrettably no such illustrated corpus is available, but one would be of great value as a working aid alongside comparative material. An even more serious deficiency is the lack of a work analysing the anatomical features that will separate the species from their different bones. At the moment, without a large comparative collection available, one is often left in doubt whether observed features are truly species characteristics or just chance variations. This problem is especially acute with the separation of deer bones from those of domestic stock.

Anyone who has worked with comparative material to hand will know the inadequacy of any existing textbook for the identification of fragments, even where this has been well annotated. The acquisition of comparative specimens must be the first aim of any student of bones. On bone courses we allow our students to explore the problem and all find that a textbook is of little help, so beware of the person with a textbook giving spot judgements. If they are good enough to give spot judgements they should not need a text book!

Identification is absolutely fundamental to all that will follow and must therefore be of the highest standard. It is a straightforward matter if common sense is used but also requires the exercise of skill and careful judgement. If one is not certain of something it is only right that it should be referred for a second opinion. The identification of the specimen, except where its nature permits two differing views to be honestly held as with dogs and foxes or wolves, is the last thing that should ever be called into question. There is the danger that in an attempt to solve the bone problem on excavations people not properly trained will be turned loose on the material. It will be a tragedy if this is allowed to happen and could seriously undermine bone studies by propagating false information and hypotheses.

Comparative Osteological Collections

We have seen in the previous chapter the need for comparative material in making identifications of the very fragmentary bones normally found on excavations and for long series of specimens to study specific and individual variation. It is clear that the osteo-archaeologist must possess his own comparative collection for working purposes and must also have access to more extensive series of these species. The basic requirement will be for a skeleton of each of the animals likely to be encountered on the sites with which he will be dealing. The cost of obtaining these skeletons is so formidable that it will generally be out of the question to purchase these as prepared skeletons and they will generally have to be prepared by the individual concerned. This is quite a difficult task when one is dealing with such animals as horses and cattle. As possession of the necessary skeletons is going to be a prerequisite for working on the archaeological material potential investigators may well be deterred by the lack of facilities for dealing with such large animals, and one must ask whether there is not perhaps an alternative. It is my view that this work should not be hampered by the limitations of personal resources but that our provincial museums should be brought into this scheme. The arguments for bone collections in provincial museums have been put by the writer

(Chaplin, 1967) and it is envisaged that provincial and regional museums will play an important part in the regional organisation of bone studies in Britain. It is suggested that key regional museums, if not all suitable museums, should hold a duplicated verified collection of skeletons of both wild and domestic animals for general identification purposes and that part of this collection should be available on loan to approved osteo-archaeologists to tide them over until they have sufficient of their own comparative specimens. The museums would not necessarily purchase the specimens ready prepared but would preferably prepare their own carcasses. Apart from the initial cost of about £20 to buy necessary equipment and chemicals and the cost of the carcasses (if necessary) a horse or ox skeleton can be prepared for under ten pounds instead of the hundreds of pounds for such from commercial sources. Most museums would agree that they need such a collection for their own purposes and are prepared to buy these commercially. Even a pessimistic assessment would suggest that by preparing their own carcasses for the same outlay at least duplicate specimens ought to be available. It is envisaged that in addition to the general collection each museum would have a more extensive specialised collection concentrating on a single group of animals or a particular species. In this way all the groups concerned would be covered in depth so that when problems arose these could be referred to the special collection concerned.

We are at the present moment in the position where very few museums have a good or reliable collection of bones suitable for identification purposes. It is essential that the whole range of animals likely to be encountered should be available for comparison when seeking an identification since missing species tend to be overlooked. It is of course fundamental that the comparative collection should be correctly identified. The only way to be absolutely certain of this is for the skeleton to be prepared from a known animal. It is necessary that our key comparative collections should be formed in this way. Old material cannot of course be ignored but so many bones in our museums are misidentified either initially or through misplacing of labels that one must be very cautious of using the specimens that are available and also of attempting to identify bones of unknown origin in museums. Although these may, for example, resemble British species they may well be from an exotic. It is for this reason that I have stressed above the need for the initial use of verified museum collections.

In order to establish a comparative collection it is necessary, bearing in mind the period of time with which one is dealing and also the geographical area, to draw up a list of those species required and to define priorities for their acquisition. In Table 1 I have drawn up a

Table 1. Composition of the British post-Pleistocene Mammalian fauna by Orders

Order	Examples	No. of species
Insectivora.	Hedgehog, Mole, Shrews.	7
Chiroptera.	Bats.	15
Carnivora.	(Terrestrial) Dog, Stoat, Otter.	14
Carnivora.	(Aquatic) Seals and Walrus.	7
Artiodactyla.	Pig, Deer, Cattle, Sheep, Goat.	14
Perissodactyla.	Horses, Ass, Donkey.	3
Lagomorpha.	Hares and Rabbit.	3
Rodentia.	Squirrels, Beaver, Mice, Voles.	17
Primates.	Man.	1
Cetacea.	Whales, Dolphins and Porpoises.	23
	TOTAL	104

This total differs from that of Corbet (1964) as it includes both domestic forms and their wild ancestors as well as introduced forms of non-British origin.

list of the orders of post-Pleistocene mammalia recorded from the British Isles in both terrestrial and coastal habitats. Figure 6 shows the relative frequency of excavations concerned with the different archaeological periods in Britain during 1970. From this it is possible to estimate the relative quantity of bones from the different periods and knowing the species likely to be involved we can assign general priorities for the acquisition of particular species. In Britain most of the material falls in the post-Neolithic period, much of it in the historical period. Therefore our main concern will be with domestic species closely followed by the large wild ungulates. On this assumption we can begin to evaluate the needs of persons working on post-Neolithic sites in the country and also for Mesolithic sites. The same principles may be applied to other periods and areas.

It is the large ungulates (hoofed animals) which must almost always have formed the bulk of the animal protein consumed by

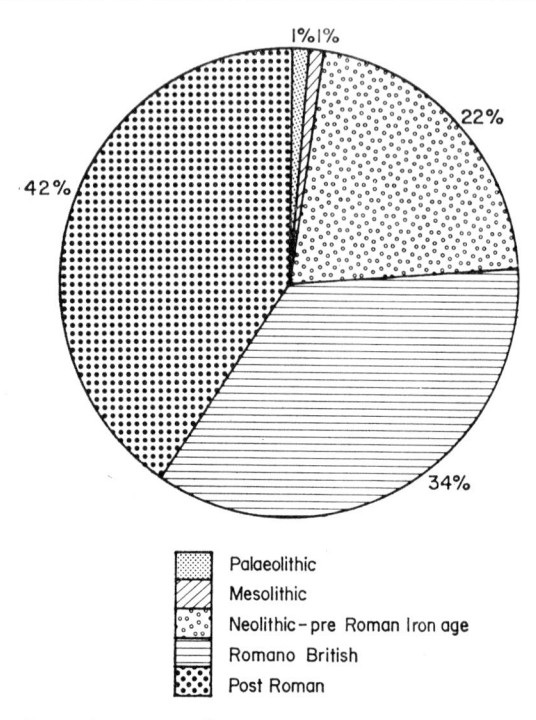

Figure 6 The relative frequency of excavations on sites of different ages in Britain during 1970.

ancient man except in coastal environments. Consideration of the size and organisation of the likely social unit combined with an estimate of the ratio *time spent in capturing/amount of meat captured*, supports this view. One must not, however, ignore the ecology of the area in question and the relative abundance of large or small game in the area involved. Some habitats would not provide a choice between large and small game and also technical achievement may not have been sufficiently evolved or adapted to permit such a choice. Further, one may possibly encounter cultural inertia in a rapidly changing environment such as that of North-West Europe at the end of the last glaciation. Nor should we ignore the unknown factors influencing human spatial distributions and hunting areas. The work of the archaeologist often helps but little with these problems since we are viewing the environment through a cultural filter, and contemporary natural accumulations may be overlooked because of the lack of human associations. Whilst these problems must concern us they should not be allowed to detract from the general principles involved.

Considering first of all sites from the agricultural period, we shall need to consider both domestic and wild stock, for whilst domestic stock may be more available, the taking of wild species in numbers sufficient to be of significance in the diet may be a necessary part of crop protection. Even on many historic sites where wild species are often of no significance, examples of wild species will still be required. Although we may know that cattle, sheep and pigs were probably of overwhelming importance we would still require examples of other species known to be present during the period so that they could be certainly eliminated. We need both those species which are of frequent occurrence and also those with which they could be confused. Thus of the larger British ungulates we would require horse, cattle and red deer skeletons to be sure which species was involved.

On this basis Table 2 has been drawn up in size groups which show both the comparative species desired and those species which will need to be considered in a particular size range. The great bulk of the bones from archaeological sites comes from animals ranging in size from the horse to the sheep; a small proportion is from cats and dogs and a varying proportion consists of smaller bones deriving principally from domestic and wild birds, fish, and medium to very small wild mammals. Generally this proportion is very low. It will be

Table 2. Comparative size of British mammals and birds based on principal bones

LARGEST

Cattle (*Bos* sp.).
Horse including Ass and Donkey (*Equus* sp.).
Red deer (*Cervus elaphus*).
Elk (*Alces alces*).

Fallow deer (*Dama dama*).
Roe deer (*Capreolus capreolus*).
Reindeer (*Rangifer tarandus*).
Pig (*Sus* sp.).

Man (*Homo sapiens*).
Brown bear (*Ursus arctos*).

Walrus (*Odobenus rosmarus*).
Most Cetacea.

Largest birds.

MEDIUM

Sheep (*Ovis* sp.).
Goat (*Capra* sp.).
Roe deer (*Capreolus capreolus*).
Wolf/dog (*Canis* sp.).
Fox (*Vulpes* sp. and *Alopex* sp.).
Badger (*Meles meles*).
Otter (*Lutra lutra*).
Wild and domestic cat (*Felis* sp.).
Indigenous and vagrant seals (six species).

Hares (*Lepus* sp.).
Rabbit (*Oryctolagus cuniculus*).
Beaver (*Castor fiber*).

Small and immature Cetacea.
Medium-large birds.
The young of largest group above.

SMALL

Hedgehog (*Erinaceus europaeus*).
Rabbit (*Oryctolagus cuniculus*).
Pine marten (*Martes martes*).
Stoat (*Mustela erminea*).
Weasel (*Mustela nivalis*).
Polecat (and ferret), (*Mustela putorius*).
Red squirrel (*Sciurus vulgaris*).
Beaver (*Castor fiber*).
Black rat (*Rattus rattus*).
Brown rat (*Rattus norvegicus*).
Medium size birds.
Young of medium division above.
Larger bats.

LESSER

The balance of the British list including shrews, bats, mice, voles and the young of the small group above.
Small birds.
Some fish bones and those of reptiles and amphibians may also be of similar size.

This list reflects the size of the bones which does not always reflect absolute differences in height, length or weight of the animal. It is therefore only an approximate guide to animals having bones of a similar order of magnitude to each other.

seen that in groups a and b in Table 2 there are only fourteen species similar in size to, or larger than a sheep. For animals smaller than the sheep there are more species involved and in attempting to carry comparative material to deal only with the mammals about eighty

3. BONE IDENTIFICATION AND REFERENCE COLLECTIONS

species will be required. However, these bones are easily picked out by size and it is a simple matter to save them up and then take them along to a museum collection and identify them there. With the larger species which comprise the bulk of the material being studied one requires to have the comparative material to hand. Accordingly, the species that will need to be acquired at the outset are cattle, horse, red deer, pig, sheep, goat, fallow deer, roe deer and dog. This list omits certain species given in Table 2. A horse skeleton will serve for the separation of asses and donkeys from other species, separation of these from each other being a much more difficult problem. A domestic bovine will also serve for the anatomical separation of wild cattle from other species. A wild or a domestic pig can be used for the pigs. The brown bear is so different from other animals that one is immediately alerted and bear bones will not be overlooked. The reindeer is perhaps unlikely to be found except on a few sites, it cannot therefore have any claim to a high priority on our list. Further it is not a species that is easily obtainable in this country. The reindeer skeleton does of course follow the cervine pattern and the opportunity should be taken of comparing it with a fallow deer, which is of comparable size. Failure to obtain good agreement by comparison with a fallow deer skeleton should be sufficient to alert the worker to the possible presence of reindeer.

Of the medium sized animals, a dog and a cat skeleton will be of great assistance. It is difficult to assign any priorities to the wild species included in this category. Certainly all may be found on excavations but it is only the hares and rabbits that are of widespread significance in most excavations. The rabbit was probably first introduced to Britain by the Romans. The most important introductions came in Norman times (eleventh and twelfth century). It is a significant food animal on many post-medieval sites and because of its burrowing habits may be found in strata of any age. In these cases, and also with the smaller animals such as mice, voles, shrews, etc, it seems preferable to sort them and then make use of an official collection for taking them down to species. Since a great deal of the mammalian component of this sorted material is likely to be from hares or rabbits it may save a journey if these specimens are to hand. They are so easily obtainable and prepared, as well as making good eating, that no collection should be without one. For those who prefer not to eat their comparative specimens it is worth noting that hare and rabbit skeletons can often be picked up in the countryside and are readily distinguishable if the skull is to hand; failing this there are a number of diagnostic anatomical features of the limbs which along with size may be used to separate the two.

In the case of a mesolithic site it may be desirable to have access to a wider spectrum of species. It is unlikely that anyone will be able to specialise solely in these sites and therefore the amassing of a special collection is not of real importance. The basic collection noted above will very likely deal with those species found, which usually supplied over 90% of the meat available. The remaining 10% or so will probably be made up of perhaps twenty of thirty species of mammals, birds and fish. Thus 10% of material will require a comprehensive collection of specimens and should not be attempted by anyone who does not have access to one of the major comparative collections.

Coastal sites may reasonably be expected to produce similar material to those inland except that remains of aquatic mammals (whales, dolphins and seals) may be found. In general these derive from strandings such as occur at the present day or, as in the case of seals, hunting for the pelts and dogmeat, perhaps.

Until now, the bones of vertebrates other than mammals have received little attention. This is quite deliberate since in general they are of little economic importance on sites. Birds, reptiles, amphibians and fish do, however, have a great interest to both the archaeologist and zoologist. Although these bones have acquired an even greater mystique than mammal bones, their identification and description in no way differs from what has already been said. It is rather the fact that so many species are possibly involved (over 500 birds, and 400 fish) and that few museums have any reasonable collections of fish or bird skeletons. It is clear that a collection of this scope is not easily obtained and most animal archaeologists will perhaps confine their collecting to a few common species. There is, however, a good case to be made for one or two individuals to specialise in these groups on account of their size and to act as a bird or fish consultant for other specialists. It is here that our regional, and in particular our national museum collections, should help us on an international basis.

Having defined our priorities in the way of species for our immediate working collections, and for general collections in museums we may consider how the specialised collections might be established.

The Specialised Osteological Collection

It is essential in any investigation that sufficient information should be available to enable firm conclusions to be drawn. In relation to the work of the specialists we have already mentioned the importance of ascertaining the diagnostic value of anatomical details. Equally it is essential to know the range of variation of such things as the length or width of a bone and the age at which an epiphysis fuses with the shaft. Our ability to deal with species from archaeological sites will be seriously hampered unless such things as these are understood for populations of known character.

It is therefore very regrettable to find that this sort of information is rarely available. At the present moment many profitable lines of enquiry are left open because the necessary information is not to hand and time cannot be spared to undertake the necessary work. Much of the information likely to be required by the specialist falls outside the scope of many modern biological studies so that it will be supplied only incidentally, if at all.

The question is, of course, how will this information become available. No doubt in the future there will be opportunities for carrying out investigations of these problems in relation to archaeological material. It is, however, unfortunate that a great deal of skeletal material goes to waste from specimens collected for other purposes. Such a waste of biological material cannot be viewed without concern by those concerned in any way with wildlife. It follows that the logical source of the skeletal material will be from existing programmes of work. There are, however, more scientific grounds for proposing such an arrangement. In obtaining skeletal material for our studies we aim at obtaining a true cross section of the population or species with which we are concerned. This is by no means a simple matter since our sample must be composed of different age classes for each sex, etc, and will probably consist of several hundred specimens according to the number of significant components of the population. At the outset of an investigation it may be far from clear what other biological factors like diseases or parasites may influence the primary subject of study. We are dealing with a complex biological and social structure, living animals and not inanimate objects. If we ignore the other aspects of the animals it may seriously undermine the value of the work. For example, metacarpal bones collected from a fallow deer can only be used as examples of fallow deer metacarpals. The same bones collected in regard to the sex of the animal enables the specimens to be also used in a study of sexual dimorphism. With age, weight, etc, also recorded further lines of profitable enquiry are available without the difficult task of collecting and preparing further specimens.

The person best able to collect the adequate samples required to obtain the information sought by the specialist is a zoologist undertaking work on the species. How one taps into such a source of supply depends on the individual or, preferably, the museum. It is my hope that it will be possible for regional bone centres to follow through a major study of the biology of a particular species. That this is not beyond the means of any regional museum is shown by some of the excellent archaeological and less often the biological papers produced by museum staffs in the provinces. Normally one would expect that such work would receive substantial grants from outside bodies. If such a national project were to be taken up it should enable an adequate treatment of all the major species to be produced within a few years.

As well as obtaining an insight into the nature of the animal as determinable from the bones, such a project opens up new vistas for the bone specialist as well as the zoologist and ecologist. Properly executed these studies will tell us about the living animal, its likes and dislikes, habits, etc, which will enable us, often for the first time, to begin to think in terms of animals, flocks and herds and not just of the semi-permanent framework of a late corpse.

The greatest danger that faces the osteo-archaeologist, and indeed the archaeologist and zoologist also, is that they should become obsessed with the detail and forget the totality of the organism and environment. In the case of the bones it must be remembered throughout that the detailed scrutiny of bones is but the *modus operandi* for a study of the dynamic complex of man, animal and environment. Bones were not often the concern of ancient man but the pelt, meat and milk were, as were also the habitats in which the animals could be found.

The Preparation of Comparative Material

When handling large numbers of both small and large animals, it is necessary to process this material in the shortest possible time. When

3. BONE IDENTIFICATION AND REFERENCE COLLECTIONS

we first began preparing large animals, in particular, fallow deer, it would take a week to convert the carcass to a skeleton. We can now, if we so wish, receive a carcass in the morning and have it finished by the evening; thus labour costs are reduced to a minimum. This has been made possible by the purchase of a 10 gallon domestic electric wash boiler and the use of sodium perborate as both a bleaching and degreasing agent. It should be emphasised that the standard of preparation is not as good as that of commercial preparators who generally use maceration or similar techniques as well as separate degreasing and bleaching processes. However, these processes take several months and the advantage of rapid preparation plus the fact that the results are more than adequate *for our purposes*, outweighs this point.

The preparation procedure for large carcasses is as follows:

To hang the animal for skinning and gutting make a slit between the tendons and the bone at the back of the hock joint in both hind legs and slide a metal hanger or wooden substitute through the slits, ensuring that it is long enough to allow the legs to spread apart. If wood is used drive in two nails outside the legs to prevent a leg slipping off. Pay particular attention to this as it can be dangerous if the carcass slips whilst cutting. Hang the carcass from a suitable hook, beam or branch of a tree. With an appropriate knife make a slit in the skin from the knee along the inside of each leg to the crutch and continue this down the midline of the body to the lips. Make a slit from the breast bone to the wrist joint of each front leg. Remove the skin using as far as possible only the hand to separate the skin from the underlying muscle. The skin should be preserved by pinning it out and covering the flesh side with borax, salt, etc. Slit open the abdomen with care to avoid puncturing the gut. Sever the rectum and the oesophagus where it passes through the diaphragm. The gut and stomach will now fall clear. Remove the contents of the chest cavity (heart and lungs). Strip the major muscles (meat) from the bones, do not be fussy about this, just remove the main masses of tissue. Joint the carcass into suitable sized portions for the boiler beginning at the head. The head is best removed first. The neck is generally cut into two sections, the ribs and lumbar region similarly. The pelvic girdle is left whole, the femora being separated at the acetabulum. The long bones are usually separated at the joints or the limb simply folded depending on the size of the boiler in use. The parts of the skeleton are then packed in a rigid wire basket which stands in the boiler. The boiler is filled with *cold* water and brought to the boil which takes about an hour. The boiler is then turned to low and allowed to simmer gently for a varying length of time. Two hours is enough for medium sized animals but it is desirable that any animal should be investigated after this time and such flesh as can easily be removed should be stripped and, if necessary, the bones be returned for further simmering. The key to success is gentle simmering for easy removal of the meat. Rapid boiling only toughens it. The most resistant parts are usually the joints and one must take care to see that other bones are not destroyed whilst softening the gristle. It may take four or five hours for some really tough joints to soften but as little as half an hour for the ribs and scapulae to become workable. It is essential to avoid overcooking otherwise the bone will disintegrate. Once the flesh has been removed the boiler is filled with clean water. Two holes about a quarter of an inch in diameter are drilled through the shafts of the long bones so that the marrow cavity is open. The bones are replaced in the basket in the boiler and brought to the boil. When boiling, sodium perborate is added at a strength of 3% weight/volume, which carries out a degreasing and bleaching operation. There is a tendency for the water to foam up very rapidly when the perborate is added and when a considerable quantity is to be used for the operation it is recommended that it be added in several small quantities otherwise the boiling over wastes much of the effect of the perborate.

The specimens are left in the perborate for at least an hour and then washed in warm water. The best results are obtained if the bones are left in the perborate solution until it is cold. Perborate also softens flesh and ligaments so that it has considerable value as a finishing agent as an alternative to excessive boiling. It should be noted, however, that its effect is confined to within about a quarter of an inch of the surface of flesh which it jellifies.

An alternative to the use of perborate, and which with some species or bones may have to be performed in addition, is separate degreasing and bleaching, though it does not matter which comes first. Removal of the grease can be effected by any of the customary degreasing agents such as chloroform, benzene, petrol, etc; all of these, however, are dangerous to the operator, since they have explosive or anaesthetic properties. The normal commercial degreasing plant uses trichlorethylene which, whilst poisonous, is not inflammable and with common sense may be handled in safety. As an alternative to these the bones may be soaked in cold dilute ammonia. Bleaching is normally done with 20 vol. hydrogen peroxide. The bones are placed in the solution and left until they have whitened sufficiently.

This technique may be used on any animal down to about the size of a hare. Animals the size of a hare down to that of a squirrel are best dealt with in a basket of fine mesh or a bucket to avoid loss of the smaller bones. For animals below this size cooking is not recommended because of the difficulty of removing the flesh without damage to the bones.

The alternatives to cooking are either maceration or use of carnivorous invertebrates. Maceration simply consists of allowing the flesh to rot away with or without the addition of an enzyme. The results are excellent and this is the ideal way to prepare skeletons as it involves little work or risk of damage or distortion by cooking. It does, however, presuppose possession of suitable facilities if one is considering large animals. For the smaller animals, however, it is not so objectionable since the smell can be contained by the use of a sealed jar. An animal to be prepared is skinned and roughly defleshed. A plastic label with full details of the animal is placed inside the jar with the specimen which is covered with water, the cap is screwed home, the jar labelled externally and then left for as long as necessary. The skeleton is ready when the separate bones are clearly visible. The stinking fluid is decanted off through a sieve and the specimen washed. This is a convenient point at which to add the ammonia for degreasing and subsequently the peroxide for bleaching. The bones are then picked out and placed on dark coloured blotting paper (for visibility) to dry.

As an alternative to maceration one may use various invertebrates to eat away the meat. This does, however, mean maintaining a colony of them. The principal species used are *Dermestes maculatus* (a beetle), cockroaches, or ants, of which the *Dermestes* are the most popular. These, in a reasonable colony, will strip a mouse carcass overnight leaving a semi-articulated skeleton which is bleached and degreased in the usual fashion.

As an alternative to controlled maceration or animal cleaning a carcass may be left exposed for nature to take its course. Unfortunately it is not unknown for nature to take the carcass with it! If a carcass is to be left exposed in this way it is essential that it should be protected against carnivorous animals such as foxes, gnawing rodents, and well hidden from predatory personages. It may seem logical, perhaps, to bury the carcass; in my experience, however, this is generally unsatisfactory as the carcass takes a long time to rot down especially in soils that are poorly aerated (like clay), and considerable staining of the bones usually results. If burial is the only alternative an evisceration and rough defleshing is advisable and the carcass should be buried in sand.

It must be emphasised that very great care should be taken when handling any animal especially where this is known to be diseased. When handling putrid material even greater care should be observed. It is strongly advised that before undertaking any preparation the reader should consult the article by McDiarmid (1966) on precautions to be observed when handling animals. Similarly, anyone handling archaeological material should ensure that such things as tetanus protection are still effective.

4
Techniques for the Study of Site Collections

Planning of the Site Study

In the previous section we have considered the general problems that face the animal specialist and have dealt with the way in which the bones should be collected and delivered to the specialist for examination. We have not so far considered what the specialist will do with this material. This section of the book is therefore concerned with the various studies that the specialist will make of the bones and how he will go about these. It is concerned only with the problems associated with the collection of the evidence; interpretation is dealt with in a later section. It is essential that the distinction between the evidence and its interpretation should be borne in mind throughout. The primary task of the animal specialist is to collect and record the evidence. Any interpretation of the evidence is personal.

The present chapter is concerned with the basic planning of the study, firstly with the definition of the groups of bone that will be studied and secondly with the examinations to be carried out.

The Cultural Context

The animal specailist can only work within the limits of the information supplied by the excavator, supplemented by his knowledge on more general matters. It is essential, therefore, that the closest collaboration should have been established between the bone specialist, the excavator and other specialists at the outset. For the same reasons it is now essential that with a much better knowledge of the site, the material and the problems involved, the excavator and specialist should plan the general framework of the bone study.

One of the most important differences between the work of the animal specialist and a person who merely lists the identifications made is the study of the bones in their archaeological context. In the past this has often been no more than a general dating, e.g. Claudian or Neolithic. This, however, is not enough. Whether we are referring to the total animal or only to a joint of meat the term context refers to the total environment of the animal, the physical factors such as weather, shelter, food supply, husbandry practices, and in reference to joints particularly to social and culinary practices. Because the evidence on environment in its widest sense can be no more precise than the excavator's records permit, the evidence relating to environment is hereafter referred to as the archaeological context.

In general it is unusual to find that the contexts are so limited that all the material may be treated as a single group. Accordingly, it is necessary before any examination is begun, that the excavator and specialist should get together to sort out the bones which may be studied as a single group.

As an example of the principles involved it will be useful to examine a study of the bones from a Roman villa. On such a site an obvious approach is to relate the bone evidence to the phases recognised. It must be remembered, however, that what are generally referred to as phases or periods by archaeologists are usually structural or artefactual phases and these will not necessarily correspond to economic ones. However, our archaeological context is that of these phases or periods. The bones are more likely to have been recorded structure by structure or square by square of the excavation than by phase and in some cases will be individually recorded. The excavator's individual squares are likely to create purely artificial subdivisions of the material, the perpetuation of which will only impede a rational study of the bones.

The contents of particular structures such as rooms or pits do, however, possess a significant degree of unity and deserve to be studied in relation to the pit or room concerned. Presented as a structure by structure analysis the results of the examination throw little light on the general picture of the economy of the villa but this approach provides extremely valuable information about areal specialisation of activity or use which would be lost in a study using the period as the smallest division. The detailed recording of bone fragments can be of particular value on sites that were rapidly abandoned or were overcome by a natural catastrophe (e.g. Pompeii and Herculaneum). Schmid (1967) was able to reconstruct a Roman kitchen by detailed recording of the bone and other finds. Further, the contents of a pit or a room will generally be much more closely dated and represent a far shorter period of time than the general phase context. Thus these particular deposits provide key dated specimens and associated groupings of bones whose importance and interest can only increase as bone studies advance.

One of the aims of the specialist will be to gather as much material as possible into a study group so that sufficient specimens will be available for a sound statistical analysis. On the other hand, as has been seen above and is shown also below, smaller groups may be necessary to pick out features of particular archaeological importance. From these conflicting aims it is possible to define a satisfactory working procedure. After careful consideration and a preliminary look at the bones the smallest significant working units are defined. These are then grouped on paper into larger contexts which will form the broadest working groups. Dealing with our Roman villa site these could be as follows:

> The largest significant contexts are each phase of occupation of the villa, periods I–VIII.
> The smallest working groups are
>
> Period I   Flavian levels in each of rooms 1–5
> (Flavian)
>   Each pit of nos. 1–7
>
> Period II   Antonine levels in each of rooms 1–5
> (Antonine)
>   Room 7 layer 5
>   Each of Pits 10, 11 and 13
>
> and so on for each phase.

The primary record and analysis of the bones are made for each of these contexts. To obtain the overall picture of the phase the results from each context are amalgamated. This can be done in two ways, either on paper by summing up the results or by physically amalgamating the material and reworking it. The former method is both sound and unsound for quantitative study depending on the nature of the contexts and duration of the occupance or accumulation. We may reasonably assume, for example, that of a dozen rubbish pits associated with a particular phase only one would be likely to be in use at any one time. Thus bones from the same animal are unlikely to be found in any more than two pits and the great majority will be found in only one. In this case it would be more accurate to sum the results from each pit to give the broader context. On the other hand if a large midden is being examined or several areas of rubbish accumulation co-existed then bones from the same animal are likely to be found in each deposit. To sum the number of individuals from each deposit would then seriously distort the number of animals involved. In cases like this it is clearly essential that the material should be physically amalgamated as an analysis of the amalgamated material would then show by inspection which bones from different deposits probably came from the same animal. It may be suggested that this can be done on paper. Certainly adjustments can be made but they are so crude as to be almost valueless guesses. One can compare a number of measurements and on this basis decide that the bones could be from the same animals. Inspection of the specimens often shows them to be quite distinct for the nuances of anatomical details are far subtler than measurements made with a pair of callipers! Equally, many fragments such as the central portions of the radius of a sheep defy attempts to match them on paper with loose ends from other deposits. Examination, however, may show that the pieces could not fit and it is my experience that often as much as 50% of the total minimum numbers of animals present is made up of small fragments of bone which scrutiny has shown cannot belong to the same animals as ends with pieces of shaft attached.

The rule is of course that one should do nothing irrevocable. Since physical amalgamation is essential for a proper study of the wider context each piece of bone will have to be marked with its reference so that later it may be retrieved and the original sub-group reformed.

Topic Definition and the Examinations to be Made

Following the definition of the contexts it will be necessary to decide which topics are going to be studied. Although these may vary from site to site and with different excavators it is my view that there are certain topics for which it is essential to record the evidence and would regard these topics as basic to the study of almost any group of bones from an archaeological site. There are, of course, many other topics about which information might be collected but these are not universally relevant and it must be up to the excavator to bring their relevance to the attention of the animal specialist so that he may devise methods of collecting the necessary information. Topics relevant to early domestication are discussed by Chaplin (1969). I would stress that from the outset the excavator must collaborate with the specialist if the maximum amount of information is to be obtained from the bones. The specialist will of course be familiar with the general run of evidence that may be obtained from the bones. It is, however, the particular local problems on which often only the excavator can provide guidance, and without this guidance and thinking at the planning stage of the examination the routine analyses of the specialist can have only a limited relevance to the site in question. It has been my experience that the excavator often has difficulty in posing any but the most general questions and often appears unsure of the relevance of the bone evidence to the problems of the site. This hesitancy is often due to a lack of appreciation of both the possibilities and the limitations of the bone evidence. I hope this section will enable the excavator to see the range of questions that may legitimately be posed and also to understand the limitations of the evidence. At the same time I hope it will provide the animal specialist with an understanding of the basic information that is required from the archaeological study which will enable him to present data which are relevant to the wider archaeological study and which go beyond a crude list of what was found. It is the breakdown of communication and lack of thought on the part of all concerned that has led to the publication of lists of crude identifications of casually collected bones which often provide no useful information and in consequence have tended to place both bone studies and other specialist reports appended to excavation texts in some disrepute. All too often the excavator has accepted a specialist report unquestioningly whether or not it was relevant or intelligible. Both excavator and specialist must accept responsibility for this and one hopes that a more critical and a more liberal approach will prevail in future.

The methods suggested below are regarded as minimum statements of the evidence which may be obtained from the bones found on excavations. It is for the excavator to ask for the information that he requires and his responsibility to see that it is obtained and published. If, as a result of failure to ask for it, the evidence is missing, it is the excavator who must accept responsibility and not the specialist. If the specialist disagrees with the particular method asked for, e.g. whether to use fragments or minimum numbers to quantify the animals present, the excavator must decide whether the alternative offered is accurate and reliable, and will give the information required.

In the chapters that follow in this section I have considered the reasons behind and the problems connected with collecting the evidence about the six topics that I consider to be fundamental to any study of animal remains from an archaeological site. The six topics all provide data fundamental to an understanding of the significance of the animals. Table 3a lists some topics which are

Table 3a. The relationship between topics of interest and the studies applied to bones from archaeological sites

(a) Selected topics.
1. Diet in general.
2. Species utilised.
3. Quantity of meat represented.
4. Food availability, preferences and restrictions.
5. Potential availability of animal by-products such as milk, cheese and skins.
6. Source of supply of meat.
7. Techniques of butchery.
8. Methods of marketing.
9. Ancillary "industry", e.g. horn, hides, glue.
10. Social implications of the ratio of wild to domestic animals.
11. Rural economy (both husbandry and the overall pattern of agriculture).
12. Husbandry techniques.
13. Husbandry and pastoral economy, e.g. specialisation on particular commodities or uses.
14. Livestock diseases.

probably of fundamental concern to the understanding of most sites though by no means all of them are of concern on every site. These topics about which information is desired provide a basis for the formulation of questions to be put to the animal specialist. Bone studies can provide evidence relevant to these and many other topics as well. Table 3b lists the examinations and studies which need to be made on the bones in order to deal adequately with these topics. The numbers following the description of the examination refer to the numbered topics in Table 3a and indicate that the examination is of relevance to that topic. It will be seen that practically every examination is relevant to at least six topics and that in most cases several studies are necessary to deal adequately with any single topic. Because of the interdependence of the two and because these topics are of concern to the archaeologist and others, I consider that collecting the evidence relevant to them all forms a desirable minimum requirement for any study of animal bones from excavations. Of the eleven examinations listed in Table 3b, six (4, 5, 6, 8, 9 and 10)

Table 3b. Examinations and studies available
(The number following the topic indicates that it is relevant to the topic bearing that number in Table 3a)

1. The nature of the site and the deposits from which the bones came. This is fundamental to any study and interpretation of bones from an excavation.
2. Geology and topography of the site. 2, 5, 6, 10, 11, 12, 13, 14.
3. Other biological studies (plants, mollusca, etc). 1, 2, 5, 11, 12, 13, 14.
4. The species represented. 1, 2, 3, 4, 5, 6, 9, 10, 11.
5. The minimum number of animals present. 1, 2, 3, 4, 10, 11, 12, 13.
6. The minimum number of animals determined for each of the bones of the body. 1, 3, 4, 6, 7, 8, 9.
7. Marks left by butchering. 2, 7.
8. Age of animals at death. 1, 3, 4, 5, 11, 12, 13, 14.
9. Anatomical and metrical estimation of sex. 1, 3, 4, 5, 9, 11, 12, 13, 14.
10. Size of animals. 1, 3, 4, 5, 8, 9, 10, 11, 12, 13, 14.
11. Disease, injury and malformation. 4, 5, 11, 12, 13, 14.

are fundamental and any report which fails to consider these must be regarded as incomplete. It will be noted that I have included at this point the collection of ancillary evidence which will only be utilised at the later interpretive stage. It is brought in here to emphasise the need for collecting further evidence which, were it not available, would considerably limit the way in which the evidence derived from the bones could be utilised. It will be noted that these studies often fall outside the field of the bone specialist and are in most cases the field of a comparable discipline. This is not a reason for neglecting them. Evidence of which plants were available in the area and which mollusca, parasites, etc, occurred are of very great relevance. Animal archaeology is not an isolated subject but like the animals it seeks to study it must be very much concerned with their environment—the food they ate, the rain that soaked them and the ticks that sucked their blood.

Regretably it is far from customary, especially on later sites, to collect this sort of evidence. Such a failure is very unfortunate for until this wider spectrum of information is available from many sites the potential of fringe archaeological studies such as palaeozoology and palaeobotany will never be realised. One cannot at times help but wish that the number of excavations would decrease and the range of evidence collected increase.

The way in which the evidence requirements have been drawn up are quite straightforward and it is useful to examine this as it illustrates the principles of approach which will be of value in framing and specifying the request for data relevant to other topics.

Modern animals are still broadly similar to those of ancient times and the same principles of husbandry are still relevant even if they differ in detail or emphasis. Thus, when faced with a specific question to answer (e.g. whether or not cattle were kept for milk), we can work out the information that we need to collect by reference to an existing herd kept for that purpose whose relevant parameters may be determined by inspection.

Given the question "do these bones represent animals kept for milk" we may approach the problem as follows. Our first task is to decide which population parameters are relevant to this question. As yet we do not know anything about the animals from which the bones came and so our problem is that of someone looking at a herd of cattle and asking the same question. The first parameter is clearly sex since male cattle do not give milk. If all our cattle turned out to be males or castrates then clearly the cattle were not kept for milk. If they are females then the evidence is no more than suggestive that they were kept for milk and therefore calf production as well since they could also be prime suppliers of meat. (It should be emphasised here that we are concerned with prime and not residual use.) We must seek the parameters that would distinguish between cows kept primarily for milk and calf production and those kept primarily for meat. The relevant factors here are clearly the optimum meat carcass weight age and the economic milk/calf yield age. Our second critical parameter is therefore the age range of the animals. In the milk/calf herd we would expect an age range extending into old age whilst in the meat herd we would expect the animals to be at or below the optimum meat carcass weight age which would represent a comparatively young population.

The basic parameters or topics needed for bone studies dealt with in successive chapters have been worked out by the use of these

4. TECHNIQUES FOR THE STUDY OF SITE COLLECTIONS

model populations and they will also be of great value when interpreting the findings.

Working Procedure

We now have sufficient information to proceed with the study of the remains and before we consider the exact way in which the information will be collected it will be useful to outline the schedule for collecting our basic data.

Having decided on the working groups and the parameters to be recorded each group of bones is sorted according to type of bone so that all femora, humeri, etc, are together, irrespective of species. These are then identified to species. The minimum number of animals represented by each bone of each species is then determined and on the same basis the number of fragments and their weight are also recorded. At this stage, with all the bones/animals laid out, the age criteria are recorded and any particular features such as pathological lesions, butchery marks, signs of cooking, etc, are noted. Those bones that can be measured and used for sex estimation are noted and separated, the "minimum animal" from which they came being recorded. The specimens that have been "noted" will receive further attention and are set on one side. The remaining groups are treated in the same way. All the noted material is then brought together and the particular features recorded and the bones measured, etc. The reason for leaving the description and measurement is to ensure consistency in qualitative description and to reduce the personal variability in measurement.

Variations on this procedure are of course possible but having tried them I find that all are less efficient and the results less precise because of inconsistencies due to personal bias.

Quantifying the Species

The method that is used to quantify the occurrence of the different bones or different species is of fundamental importance since these figures may be extended to calculate probable lengths of occupation or the number of people that could have been fed. In addition, these figures will be used to assess such things as the dependence on wild or domestic species, a dependence which does of course have far wider social, cultural and environmental implications.

At the present moment there are three methods which are being used separately to quantify the results. These are:

1. The "fragments" method which uses the number of fragments identified.
2. The "weight" method which uses the weight of bone.
3. The "minimum number" method which determines the absolute minimum number of animals from which the bones have come.

The methodology proposed elsewhere in this book reflects my view of the controversy that exists over the merits of these methods and this chapter is therefore also a justification of my choice of the "minimum numbers" method as a basis of a methodology for site investigations. It will be noted, however, that whilst other methods have been rejected as a basis for a full study of the bones, they do form part of this methodology; it is important that this distinction be clearly understood.

Fragments Method

This method of quantifying either the frequency of different bones or of different species uses the total number of pieces of identifiable bone to obtain this. This method has as its base the assumption that from the moment of death to the time that they reach the specialist's bench *all the individual bones of all the species* are equally affected by chance or deliberate breakage and will survive equally well the hazards of different methods of cooking, preservation in the soil, excavation and transport. These are the conditions that must be fulfilled if any credence is to be attached to results obtained by counting the number of fragments.

Accepting for the moment this hypothesis at its face value there are four problems concerned with the methodology which constantly recur where this method is used. These are:

1. The attention to be paid to bones which are complete but broken in the soil. How many fragments should this be counted as?
2. How should pieces of bone found separately, but which unite, be counted?
3. Fragments of bones not "identifiable" in themselves may become identified if they are found to unite with an identifiable

4. TECHNIQUES FOR THE STUDY OF SITE COLLECTIONS

piece. How should such fragments be counted and how far should attempts be made to match all unidentifiable fragments with others?

4. How is a complete burial (generally of a dog) to be counted? Such burials are frequenty omitted from the sums.

In answer to the first two points, I would suggest that the true number of fragments which could be identified separately is the figure required. In the third case I would also suggest that only the separately identified fragments should be counted. In the fourth problem each bone or identifiable broken piece should be counted. In this way one is able to carry out some sensible reunifications without infringing the "all-or-nothing" policy which would be necessitated by a departure from this suggestion. Departure from this "all-or-nothing" policy would mean that the definitive statement given above would be infringed since a definite bias had been introduced.

Any attempt on the part of the expert to comply with this hypothesis does of course make nonsense of the results if the requirements of the hypothesis are not fulfilled. Whilst such things as modern breakages can be recognised and compensated for it can be shown that it is highly improbable that the hypothesis is in fact true for any point in time following the animal's death. Therefore, any effort by the expert to avoid the introduction of bias to the sample is misplaced since this has already occurred and is an inherent property of the collection. The principal objections that can be made to the working hypothesis on which use of the "fragments method" is based are as follows. The first thing that happens to an animal once it has been slaughtered is that it is divided up into conveniently sized portions for storage or marketing, and subsequently into joints for consumption. The number of pieces that any given bone will be cut into pursuant to the preparation of usable joints is clearly related to the size of the animal. For example, there is much more meat around the femur of an ox than there is round that of the sheep.

It may therefore be expected that if a usable joint is about the size of a leg of lamb, the femur, in the case of sheep, will probably survive butchery and cooking, whereas that of the ox may be cut into a dozen or more pieces. In many cases the leg of beef may be boned out before the meat joints are cut and the bone chopped up to provide a meal in itself. Of these dozen pieces, perhaps five (to be conservative) will be identifiable compared to only one intact bone of the sheep. There is therefore a bias if the *species ratio* is based upon the number of fragments.

Unfortunately the number of fragments does not assist in calculating the *dietary ratio*, since the same number of diagnostic fragments (based on anatomical features) will be found irrespective of the size of the animal whether the bone was cut into ten or twenty pieces. Thus, although the number of pieces into which it is cut could reflect the order of magnitude of the quantity of meat involved, it does not do so in practice: the extra number of pieces must be overlooked because they are not specifically identifiable. The number of pieces into which any bone will be cut will of course be influenced by the size of household concerned and thus on some sites we may expect to find substantially complete bones if large portions of an animal were to be cooked whole, whilst in small households, we would expect only fragmentary bones deriving from small joints. Butchering practices in antiquity, about which we are extremely ignorant, will also influence the number of pieces produced.

It is a present-day practice with large animals to sell meat filleted so that with beef from the leg no bone is included in the joint. On subsistence sites where this was a practice we might reasonably expect to find the filleted bone on the site, though probably not intact, since it would be broken up and used for stew, etc. In a commercial economy, however, as with a butcher's shop, the bones would probably be sold for fat extraction, etc, and so for many joints of meat there would be no bone evidence to be found in the consumer's refuse. From these points it is clear that there is from the outset a bias towards particular bones and species of a sort which cannot be predicted or determined. From the hypothesis and also from the way this method has been used in a number of reports, it is implied that the number of fragments from each species is proportional to the number of animals involved, and is therefore a reflection of the frequency of the animals. This, of course, has been shown above to be false.

With meat purchased from a butcher in the form of joints, as is done today, it is extremely unlikely that any quantity of bone found in household refuse would have come from the same animals. In cases like this it might appear that the number of fragments is a much more realistic basis for estimating diet. This may be true apart from filleted meat. A more satisfactory method, however, and one which permits of the recognition of butcher's meat, is to divide the number of fragments by the number of animals represented. A low ratio then suggests butcher's meat, a high ratio, the much fuller use of individual animals. The type of bones found will refine this still further allowing one to suggest whether or not a whole carcass was bought or the animal slaughtered at the point of consumption. A further objection to the sole use of the "fragments" method is the way in which it limits the type of information that may be obtained. Reference to Table 3 and to succeeding chapters shows that many topics could not be examined if the only datum presented concerned the number of fragments found.

I have come to the conclusion that the "fragments" method is unsuitable for the comprehensive study of the bones from a site. It is, however, a method in fairly common use and one which has been more widely used in the past. It is superficially attractive in that counting fragments is both simpler and quicker than determining the minimum number of animals. Its use is, however, time completely wasted for it allows no comparisons to be made between any two sites because the bias which is certainly present cannot be detected or determined. There can be no comparison even between similar or identical cultural groups since the decisive factors are purely personal and economic. Since the majority of bone studies have a comparative purpose we must seek other methods which allow direct comparisons to be made between different sites, cultures and geographical regions.

It is regrettable, but nevertheless necessary, that we should set aside the results that have been obtained from the use of this method. It means, very often, that since the material from many of these sites has been discarded that we shall have to write them off. It is a tragedy because so few sites have had the material studied, and most of these are sites of high significance to archaeology. The lesson that is to be learnt from this is once again to do nothing irrevocable and if we learn at least to follow this precept past efforts will not have been entirely in vain.

Weight Method

In this method, the weight of bone from each species is multiplied by a factor to give the amount of meat represented per species. This could of course be either the weight of each bone type, the weight of each species or the weight of all bone.

This method assumes that there is a constant relationship (within an acceptable range of error) between the total dressed carcass weight and the weight of the component bones. Dressed weight should be used because of the diurnal variation in live weight of an animal due to the ingestion of food and the longer term variations in weight due to conditions such as pregnancy. Admittedly there is controversy over which weight should be used since the food value of an animal does depend on dietary practices and there are of course different methods of defining dressed carcass weight. These are, however, irrelevant at this point since if the method were to be in general use, standard methods, weights and definitions would be formulated as necessary.

It is clear of course that the method is mainly intended for the calculation of dietary ratios. There are many objections to its use for the determination of the species ratio including many of those mentioned in connection with the fragments method. Since it is but a partial method of study its use at present is best confined to the solution of particular problems within the limits of the restrictions or queries raised below.

For modern livestock, almost all of which is "improved" if not highly specialised, information is available on the amount of meat, waste and by-products which should be produced by specified strains of stock kept under specific conditions. Comparable figures are not of course available for ancient animals and estimates—largely guesses, however informed they may be—must be made. This deficiency would be overcome if a relationship were to be established between a bone measure, live and dressed weights and nutritional level. There is clearly a general relationship between size and weight of an animal and its bones but the very many complex factors involved have yet to be established. It would seem more likely, perhaps, that bone weight would show the closest correlation. We cannot, however, assume that weight of bone is something that remains constant once an animal achieves its full development or that different species and genotypes have the same bone/body weight ratio. The bones are formed to take an overload and the weight of bone varies according to the normal physiological demands of homeostasis.

The relationship of bone weight and body weight is not an exact one. This relationship and the nature and extent of its variation can be determined in modern livestock and by careful choice of breeds it would be possible to indicate the applicability of these ratios to ancient stock. The range of error of such ratios may be less than the errors inherent in studying archaeological material.

The relationship between the weight of meat around a bone and the weight of the bone is not the same for different bones since the strength, and therefore, perhaps, in part the weight, of a bone is related to the total force exerted upon it. This being so the meat/bone ratio must be established for each bone, at different ages of the animal and for each sex. To be strictly accurate then, an analysis based on the weight of fragments would involve determining for each piece of bone the sex and age of the animal in order to determine the amount of meat represented. Many of the bones are of course fragmentary and thus in determining the weight of meat from each species only the weight of identifiable fragments can be taken into account. It has been demonstrated in the section on the "fragments method" above how this will bias the results. It follows that the only weight which can then be validly used is the total weight of all bones which means that we cannot assess the role of individual species but only estimate the amount of meat that could have been obtained from the bones. This of course assumes that the meat/bone ratio is constant for all species, which is unproven; in the archaeological sphere it means also that on some sites, e.g. town sites, we shall not know how much meat could have been eaten but only the minimum amount of meat consumed.

Although many facts which are fundamental to the use of this method remain to be established I have little hesitation in rejecting it outright as a method for the analysis of a site. Many features of it, however, in particular the bone/meat ratios, are of fundamental importance in an evolved methodology.

Very few sites have been tackled solely by this method and it is not in general use so that there is little work to be reassessed. Like the "fragments" method sites done by this method would not be comparable.

Minimum Numbers Method

It will be noted that the fragments and weight method have as their basis certain hypotheses the validity of which has been seriously questioned. Further, each of these methods is subject to bias in a manner which cannot be determined. The use of the minimum numbers method differs from these in that, firstly, it involves no hypotheses and is purely factual. The minimum number of animals that the bones could have come from is an indisputable fact. It is, moreover, a direct measure of a number of animals involved and is an abstraction of the true number of animals involved only within fixed limits. It also involves no assumptions about differential preservation of bone which cannot be checked by an examination of the specimens or by a site inspection. It is therefore using verifiable facts throughout. The purpose of the specialist's examination is to record the basic facts and to provide a sound basis for extrapolation from and interpretation of the facts. Necessarily the interpretative side of bone studies is extremely important; without this it would be a very arid study. All extrapolation or interpretation is of course speculation, but where it is based on the minimum numbers of animals it does have a realistic and factual basic measure of the species and their use. The arguments set forth above about the weight and fragments method show that these do not provide such a basis.

Since we are dealing with specific quantities and not arbitrary measures of these quantities all sites treated on the basis of the minimum number of animals are directly comparable. Thus for the first time we have a means of comparing sites and specifying both the nature, direction and rate of change of man's manipulation of wild and domestic stock.

Determination of the Minimum Number of Animals

The minimum number of animals will serve as a basis for the study of the site and must be determined with care. For any given species the minimum number determined will be proportional to the amount of effort expended. The first step is to determine the minimum figure for each bone. At its crudest we say that if we have eight right and nine left thigh bones these must have come from at least nine animals. Equally, if we have from the right humerus eleven lower ends and fourteen top ends these will have come from at least fourteen different animals. This, however, is not necessarily a very satisfactory estimate of the true minimum, for in the latter case they might have come from twenty-five animals (11 + 14). Therefore, as a next step, each of the fragments is matched against the others in respect of age and then size. Size can be a difficult criterion to apply for the right and left bones of the animal are not always the same size.

It is largely a logical exercise to determine whether the fragments are likely to be from the same or different animals, the necessary steps are set out in the worked example below. The reader may find this easier to follow and thus gain greater confidence if he has some pieces of bone at hand. For the final figure (the grand minimum total, GMT) the age of the animals and their sizes are compared. The number of animals is obtained by inspection and the age structure of the minimum population is thereby quantified.

Sex has been deliberately excluded from this description since at present few bones may be accurately sexed, and fragments rarely so. When satisfactory criteria are available it will of course play its part in determining the minimum. It should be stressed that the minimum should only be determined from features of the bone which are a fundamental property of the animal concerned, e.g. size, shape, etc, and not from features resulting from external influences. Thus poor preservation due to excessive cooking and staining through contact with metal are not valid criteria. They may reinforce a decision taken on anatomical or dimensional grounds but do not replace it.

The following example will illustrate the method of determining the minimum number of animals present for each bone:

Example

A group of thirty pieces of bone have been identified as being sheep's tibiae.

Stage 1. The fragments are divided into bones from the right and left sides of the body. There are thirteen pieces from the left and seventeen from the right. The minimum number of animals from each side as first determined is:

LEFT There are thirteen pieces as follows:

2 largely complete, one has the proximal epiphysis fused, the other not. In both cases the distal epiphysis is fused.

7 pieces of the distal end with part of the shaft. The distal epiphysis is fused in each one.

1 piece of distal end with part of the shaft. The distal epiphysis not fused.

2 pieces of the proximal end with the proximal epiphysis not fused in both cases.

1 piece of the upper part of the shaft.

At this stage the minimum number of animals present for the left side is determined from these pieces as follows:

2 pieces largely complete	= 2 animals
8 pieces of distal end	= 8 animals
2 pieces of proximal end could belong to any of the 8 distal ends already counted, as the proximal epiphysis fuses after the distal. Therefore these	= 0 animals
1 piece of shaft could also belong to any of the distal ends. Therefore this	= 0 animals
Thus 13 pieces are from at least	10 animals

The procedure is repeated for the right side.
RIGHT There are seventeen pieces of bone as follows
   4 largely complete as follows:
   2 with shaft and distal end, proximal end incomplete, distal epiphysis fused.
   1 intact. Both proximal and distal epiphyses not fused.
   1 shaft and distal end. Distal epiphysis not fused.
   6 distal ends with part of shaft, as follows:
   4 with distal epiphysis fused.
   2 with distal epiphysis not fused.
   7 pieces shaft only.

At this stage the minimum number of animals represented by the right side is determined by inspection from these pieces as follows:

4 pieces, largely complete	= 4 animals
6 pieces of the distal end and part of the shaft	= 6 animals
7 pieces of shaft, 2 of the pieces could belong to 2 of the distal ends, but 5 cannot belong to the remainder. Therefore these	= 5 animals
Thus 17 pieces are from at least	15 animals

So far then, ten animals have been identified in the bones from the left side of the body and fifteen in those from the right. Therefore there is a minimum of fifteen animals present and a maximum of twenty-five.

It is likely, however, that some of the bones from the left side are from the same animal as those from the right. The next step is to check the two sides against each other, to find out how many

4. TECHNIQUES FOR THE STUDY OF SITE COLLECTIONS

animals are represented by both right and left tibiae. This can only be done by inspection using ageing criteria, and measurements.

Stage 2. Any animal which cannot definitely be identified as NOT *belonging to any of the animals from the opposite side is assumed to be from an animal represented on the opposite side.* This eliminates the five animals determined from the shaft pieces of the right hand side and counts them as being from the same animals as five from the left. This may conveniently be noted as follows:

Animals RIGHT   1 2 3 4 5 6 7 8 9 10 11 12 13 14 15
                      C C C C C
                      C C C C C
Animals LEFT   1 2 3 4 5 6 7 8 9 10

     C = comparable with one from opposite side.

Of those animals where the distal epiphysis is not fused, there are four animals from the right and one from the left. The distal ends are measured to see if the left is the same as any from the right.

Maximum width of distal end of immature tibiae

left	*right*
20·4 mm	21·2 mm
	21·8 mm
	21·8 mm
	23·1 mm

The measurements show that it is not likely that the left bone is from the same animal as any of the right bones, this makes five animals represented by right and left bones. These are noted as D in the record.

Animals RIGHT   1 2 3 4 5 6 7 8 9 10 11 12 13 14 15
                      C C C C C D D D
                      C C C C C D
Animals LEFT   1 2 3 4 5 6 7 8 9 10

The remaining category consists of those with the distal epiphysis fused. These are then measured and the measurements of the two sides are compared.

Maximum width of distal end of mature tibiae

left	right
27·0 mm (C)	26·6 mm
26·0 mm	27·0 mm (C)
23·5 mm (PC)	23·8 mm (PC)
28·0 mm	25·6 mm
24·5 mm	23·4 mm (C)
24·6 mm	25·8 mm
27·8 mm	
23·4 mm (C)	
24·6 mm	

From the measurements it is clear that two marked (C) and probably a third marked (PC) should be regarded as belonging to the same animal and these three are noted as Cs in the record and the remainder as Ds.

Animals RIGHT 1 2 3 4 5 6 7 8 9 10 11 12 13 14 15
               C C C C C D D D D C C C D D D
               C C C C C D C C C D
Animals LEFT   1 2 3 4 5 6 7 8 9 10

It will be seen that all the animals for the right side can be accommodated in the notation but that there is a single animal unaccounted for in the left hand series. It is remembered, however, that five of the animals from the left were previously equated with five shafts from the right in stage 2, the two could not be directly compared because no measurements could be made of the shafts. Five of the left distal ends were the animals which the right hand side was matched against. They cannot be considered as distinct because there is no measurement to compare them with, and they are left recorded as Cs. The sixth, however, has no counterpart on the right side so no. 10 left is recorded as D.

The grand minimum number of animals represented on the whole sample is obtained by counting up the number of Cs and Ds in the table.

The grand minimum number of animals, $GMT = C^t/2 + D^t$

Where C^t is the total number of Cs noted in both right and left and D^t is the total number of Ds noted in both, C^t is always an even number. In the present example

$$\text{GMT} = \frac{16}{2} + 9 = 17$$

There are then in this group of material, thirty pieces of bone which are derived from a minimum of seventeen animals. This figure may be contrasted with the number determined solely by counting the number of right and left bones and with the maximum possible obtained by summing the individual totals of right and left. Because of the care taken it is clear that the grand minimum number determined is very close to, if not the true number, of animals killed to produce these bones.

The age of these seventeen animals can now be determined. This has not been done in the example to avoid confusion because it is not easy to demonstrate this on paper. Using the following method it is, however, possible to age each of these individuals by inspection. On paper we record the fragments of bone as Cs and Ds in a table form. To determine the age of the individuals we lay out the fragments in exactly the same way on the bench. Thus at the end we have all the pieces of bone relevant to each animal laid out animal by animal and we then examine the fragments of each animal and record the criteria for ageing.

5
Age Determination from Bones

When assessing the husbandry and hunting practices of an ancient people it is essential that one should know the relative ages at which the animals were killed or died. Determining the age of the common domestic and wild species from their bones is a difficult task. In many cases there is no information at all about the criteria to be considered, and a specialist going abroad may have to carry out the fundamental work on the species he is likely to encounter before being able to start on ancient material.

The zoologist trying to estimate the age of a dead animal usually has the whole carcass for study. The specialist working on archaeological material is normally trying to estimate the age of an animal from a single fragment of bone. The minimum number of animals in each deposit is determined separately for each bone type, it being almost impossible at present to correctly match up the same animal from its different bones. In practice therefore the age of an animal is determined several times from apparently unrelated bones. The degree of precision obtained each time being that applicable to the fragment under consideration. It follows that at the interpretive stage the age structure of the sample is determined either from the criteria available for a single bone type or by summation of the criteria for several bones. The method of handling this information is demonstrated in Chapter 9. In the present chapter we are concerned with the techniques that can be used to estimate the age of an animal from its bones.

There is a variety of skeletal criteria available for age determination ranging from the fusion of the epiphyses of long bones to the eruption and wear of the teeth. The relationship of the criteria can be established by examining the skeleton of animals of known age. It is generally agreed that the age at which some of the critical events occur has changed with selective breeding. The greatest variation is thought to have occurred with the eruption of the teeth. We do not know to what extent other criteria, such as the fusion of the epiphyses differs between ancient and modern stock or whether any change has proceeded at a proportional rate for all criteria. Caution is therefore needed when comparing the ages obtained from different criteria. Complete skeletons provided a valuable check on these relationships in ancient animals.

The methods of age estimation discussed below provide either a relative or an absolute base for ageing the animal. Absolute methods, such as those which count growth increments, are recorded in appropriate units of time. Other methods which are essentially relative, e.g. the eruption of teeth or the fusion of epiphyses may also be recorded in absolute time units. This is possible because the age at which these sequential events occur has been determined by appropriate experimental methods. In the case of ancient material where the true ages cannot be determined it is permissible to record the stages present and to assign to them, with appropriate explanation, a time scale determined from modern stock. The following methods of age determination are applicable to bones from archaeological sites:

1. Tooth eruption and replacement.
2. Epiphyseal fusion.
3. Closure of cranial sutures.
4. Incremental structures.
5. Tooth wear.
6. Antler development.
7. Size and form (growth).
8. Qualitative features.

References to each technique are given at appropriate points below. However, a recent review in Russian (Klevezal and Kleinenberg, 1967) surveys techniques of age estimation in bony tissues and has now been translated by the Israel Programme for Scientific Translation and the Fisheries Research Board of Canada (translation

No. 1024, 1969). Techniques are also reviewed by Sergeant (1967), and a bibliography is given by Madsen (1967).

Tooth Eruption

Tooth eruption is only of value until the last tooth in the series comes into wear. As it happens many animals are killed before all the teeth have erupted and this method is singly the most useful of the eight methods. It is usual that an age is given to each stage in the development of the permanent dentition and is usually referred to as the age at which a particular tooth erupts. The ages at which these teeth erupt in different species are obtainable from a number of veterinary sources. Convenient summaries are those by Silver (1969) and Habermehl (1961); the latter should be studied if possible, especially in regard to variability. Examples are given which may be of relevance to a particular site. The ages given in veterinary sources usually refer to the age at which the tooth cuts the gum which is of course later than its penetration of the bone beneath and earlier than its beginning to wear. These distinctions are readily seen in archaeological material but for practical purposes the differences are usually negligible.

The order in which the teeth erupt is largely constant for a given species though minor variations may occur in local populations. However, the age at which a tooth erupts is subject to considerable variation. The main factors involved appear to be breed differences, the diet of the population, and environmental conditions. Valid figures for tooth eruption can only be derived from studies on a population of known character kept under specific conditions. The figures derived from these studies are applicable only under the same conditions for that population. Within this population there will be variations between individual animals and a figure quoted will be the mean for that population. More generalised figures applicable to that species under a wide range of conditions can be derived by extensive studies, but some loss of precision must occur and the range of possible age may be great enough to render the figure of little value for economic studies. As the character of the population and the conditions under which it was kept cannot be deduced from the bones using present techniques one is forced to accept such a wide range if absolute ages are desired. Because we cannot isolate the factors that might make for the variability of the ancient population we cannot be sure which of the series of figures available will be most applicable to the bones. We may certainly suggest that figures for more primitive breeds or those derived from early husbandries are likely to be more relevant than those from modern highly bred animals, but we must be clear that this is no more than an informed guess and remains largely unproven. It does, however, seem possible that incremental structures in the teeth of domestic animals could provide an absolute age for a tooth enabling a more appropriate chronology to be used. It would, however, be necessary to test this applicability for each site as conditions may well have varied significantly between sites in any single cultural period. It would be very misleading to talk for example in terms of Iron Age species and economies without reference to numerous examples derived from the wide range of environmental conditions and economic levels which occur in any geographical region.

The principal use of age determination will lie in an interpretation of the husbandry and economic practices from which the bones are derived. From the foregoing remarks it will be apparent that however many sensitive stages can be distinguished in a tooth succession they will be largely invalidated by the imprecision of the ages that can be assigned to them. A further limiting factor will be the variability in the date of birth of animals which, for a single population, will cover at least several weeks. Variations in the mean birth date can occur from year to year determined for most animals by the time of mating in the previous year. It would appear that in some seasonal breeding species the time of birth may be controlled by the mother in relation to environmental conditions.

In time, ways of overcoming these objections may be found. In the meantime there is no real objection, beyond the fact that it may be misleading to persons unfamiliar with the problems, to the use of absolute ages in a report, provided that certain basic scientific principles are followed. In particular the criteria for the age assigned should be clearly stated and the facts clearly separated from the hypotheses developed from them.

As has been mentioned above, the succession of the teeth is the key to the ageing of the bone. Discrepancies or variants within a population may be found and examples have been described by Ewbank *et al.* (1964) in some Iron Age sheep. Where a large sample is available the sequence of eruption should be examined. Some implications that may be derived from a study of the variability of the sequence are noted in Ewbank et al. (1964). This is primarily concerned with the open or closed nature of the breeding population under study.

Many of the problems of age determination as they relate to the indications of economic activity can be overcome by the use of group frequency analysis. In this technique all the jaws showing a similar stage of development (not age) are grouped together. The stages are then arranged according to the order in which they occur and the result is most conveniently displayed as a frequency graph which shows the stage at which animals were killed. In analysing the sheep mandibles, which are particularly frequent on a number of sites, the writer and others have found that a considerable percentage of the jaws often shows identical stages of development even when very close stages are being used. It is possible that a group such as this represents animals of the same age, perhaps those animals whose births occurred during the peak birth period and were killed at the time of peak slaughter. It should be remembered that whilst births can range over many weeks, killing can be done within a very short period if the herd is being culled, as would occur when meat was salted down for the winter. Another interpretation might be that a certain percentage of the herd was killed as each animal reached a certain size or weight. Because all the animals had a similar genetic make-up and environment the optimum size or weight corresponded with this degree of tooth development although the killing might have been spread over many months. Whatever the result, which is to be discussed in greater detail below, these very close groupings show either that killing was concentrated at certain times of the year or that it was concentrated on animals that had reached a certain degree of development determined not from their teeth but perhaps from weight. The only drawback to the depiction of these data in graphic form is that it tends to suggest that the horizontal axis is equated with a continuous time variable. It is not. The spacings are quite arbitrary. The same technique of group frequency can be used to analyse the data derived from other techniques described below.

Epiphyseal Fusion

During longitudinal growth the ends of the bone (epiphyses) are separated from the shaft by a pad of cartilage (Figs 3a and b). When growth ceases the cartilage ossifies, uniting the epiphysis with the shaft (Fig. 5). Growth ceases in different bones, and at different points in the same bone at different ages. Thus the lower epiphyses may fuse before the upper in one bone and both may fuse before either do so in another bone. The sequence of fusion in a species is regular. The sequence of fusion and the age at which it occurs can be studied from skeletons of known age or by X-ray examination of living animals. The age at which fusion occurs is, however, variable. Sexual dimorphism in bones is in part a consequence of the age of fusion. Females and castrated males have more slender bones than males as fusion is delayed. Apart from the effect of sex hormones fusion of the epiphyses may vary because of breed differences, environmental factors and individual variation. The age of fusion may also have been influenced by selective breeding. It is therefore necessary to use modern data with caution when estimating the age of ancient stock.

Epiphyseal fusion is, however, of great value in archaeological studies because most long bones give some indication of their age. A valuable relative age structure can be constructed by plotting in sequence the percentage of animals aged more than or less than a given criteria. If the sample size is adequate then a killing graph can be constructed using the difference in the percentage figure of groups in a chronological series (Figs 22–24).

The age of fusion of the epiphyses has been worked out for most domestic species but not for many of the important game animals (Habermehl, 1961; Silver, 1969). Early husbandries do not provide this information.

Suture Closure

The fusion of the separate bones of the skull, when their development is complete, is analogous to the fusion of the epiphyses. Not all the bones of the skull are necessarily affected and the details must be checked for each species. Variability in the sequence of fusion and of the age at which it occurs is affected by the same factors as have been discussed for the teeth and the epiphyses and there may be additional factors. In many species it is, however, a valuable method of approximation as it continues long after the teeth have erupted and the epiphyses have fused. It is most useful for those animals which as riding or draught animals are permitted a long life span under domestication, e.g. the horse, camel and elephant. The method

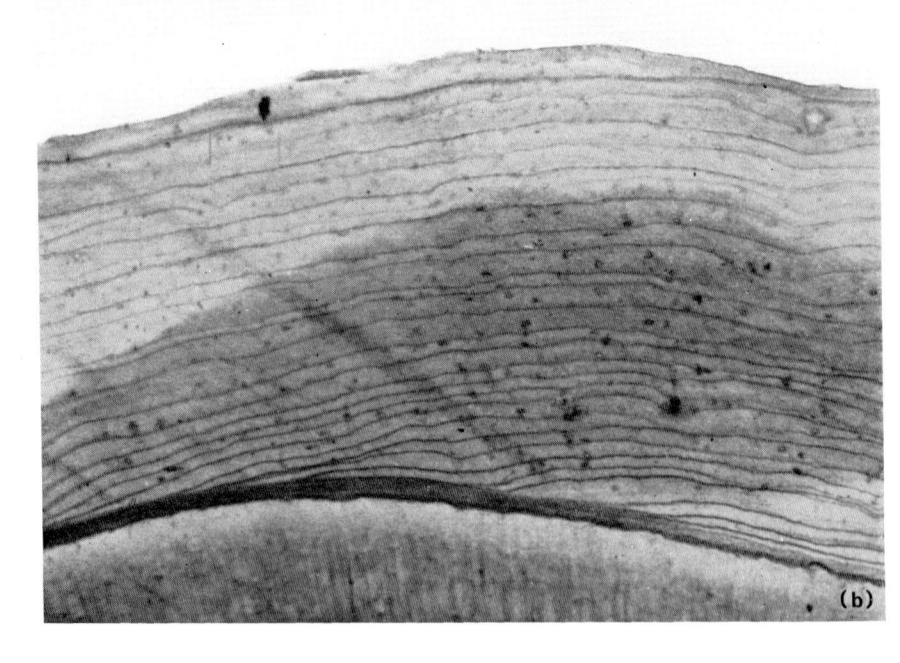

(a)

(b)

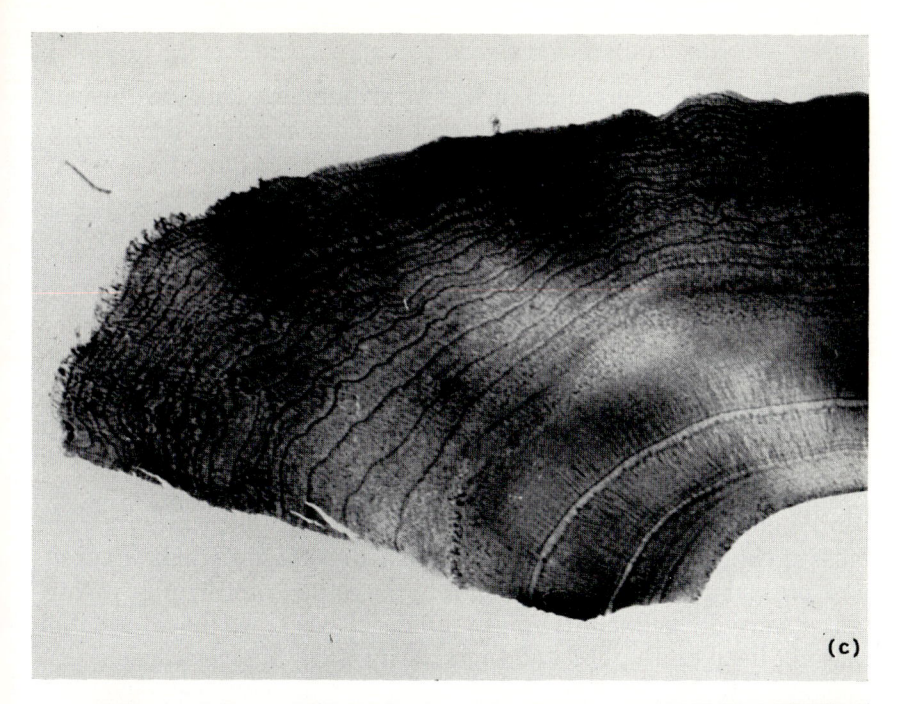

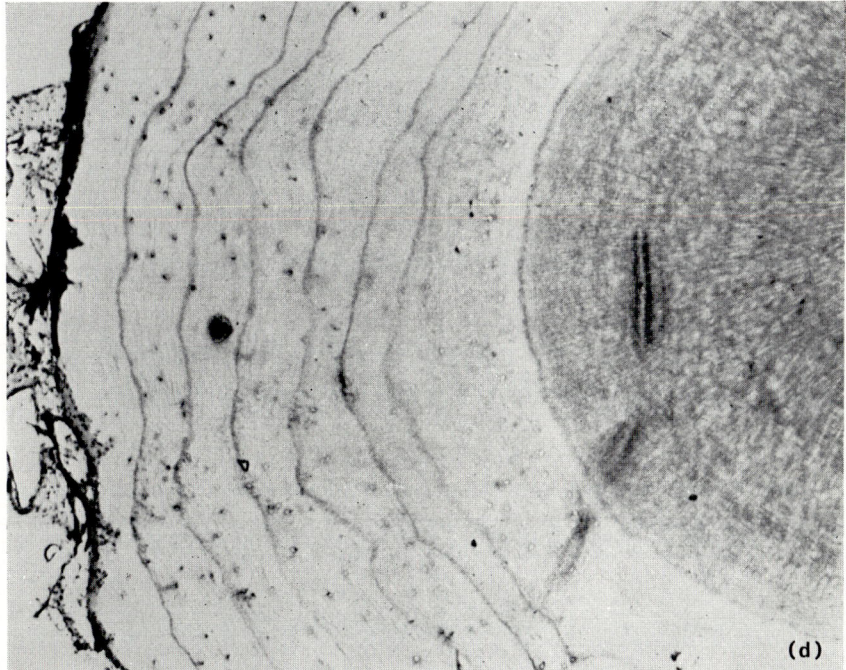

Figure 7 Incremental lines in the dental cement of (a) Cow (×8). Deciduous first incisor root; (b) Man (×45). Molar root; (c) Grizzly bear (×8). Canine root; and (d) Black tailed deer (×30). Permanent first incisor root.

is generally of lesser value in the meat animals which are usually killed at a time when other criteria are more effective. Fusions that occur in the early years should, however, not be ignored.

Incremental Structures

Periodic incremental structures have been noted in a variety of calcified tissues in many species. These increments are of considerable importance in archaeological bone as they offer the only means of determining an absolute age for the animals.

In teeth incremental structures associated with age are best known in the secondary cement deposited around the roots and below the crown of incisor and cheek teeth (Fig. 7). This cement is deposited throughout life and the rate at which it is deposited varies over the year giving a banded appearance. Although these periodic deposits are present in domestic species they have been most fully investigated in wild animals. The physiological basis for their formation has not been established and their clarity varies between species and localities. In grizzly bears, for example, they are very well developed (Fig. 7c), which may be associated with metabolic changes during hibernation. However, North American deer and most seals also have well defined layers in the dental cement (Fig. 7). From the south of England I have examined the secondary cement of the teeth of several deer species but the layers of cement are not easily counted with present techniques.

In the ungulates it seems probable that the different layers are paired, one thick and one thin layer of secondary cement being formed each year. Depending on the nature of the outer layer it is possible to tell at which season the animal was killed. As the physiological basis of these layers is not known, and they are found in diverse species, it cannot be assumed that these layers are annual in all species. In those species where they are well developed and associated with a seasonal phenomenon (e.g. hibernation in bears) the number of layers corresponds with age. In those where the layers are more weakly developed it may be that no new layers are formed in one year whilst several might form in another.

In species which hibernate or suffer a severe environmental setback during growth, these events may be marked in the bones by a line of arrested growth. Such a seasonal setback as occurs in the hedgehog mandible must be distinguished from the randomly distributed setbacks associated with disease or injury. Lines of arrested growth have been recognised in human limb bones where they are known as Harris's lines.

Incremental structures have been observed in the otoliths (ear bones) of fish and whales (Nishiwaki, Hibiya and Ohsumi 1958) and are standard techniques for age determination in many fish and cetaceans. Similar structures have been observed in fish in the opercular bone (Kyle 1926) and in the fin rays by Wells (1958). The growth rings in the scales of fish are widely used in fish management but unfortunately scales are rarely preserved on archaeological sites. Banfield (1960) has found annual structures in the pedicles of caribou (*Rangifer* sp.) and they have been noted by the writer in the pedicles of other species. Unfortunately it would appear that with age the earlier structures are obliterated.

With fresh material three techniques have been widely used to observe these structures in teeth. One method involves sawing the tooth in half and polishing one face. Gross differences in the cement are clearly visible in both ancient and modern specimens but in animals where the layers are poorly developed the cement appears homogenous. An improvement on this is obtained by grinding and polishing a thin section which is then examined under the microscope.

The most satisfactory results are obtained by fixing the tooth in formalin and decalcifying it in a dilute acid solution (e.g. 10% acetic acid). Thin sections are then cut on a freezing microtome, mounted and stained with haematoxylin or other stain. The most suitable method for examining archaeological material will depend on the nature of the preservation and will have to be determined by experience.

Useful descriptions of the techniques and their application to present species are given by Lowe (1967), Low and Cowan (1963) and Jensen and Nielsen (1968), whilst archaeological material has been examined histologically by Anderson and Jorgenson (1960).

Tooth Wear

The degree of wear exhibited by the teeth has long been used as a basis for age determination in many species. The figures derived for a local population, which may be only approximately correct, are in theory applicable only to the same population in that same environment. The principal factor, in a healthy animal, determining the rate at which the teeth are worn is the nature of the food eaten and the mineral matter attached to or contained in it. Animals grazing short vegetation on sandy soils and dusty vegetation have a much greater rate of dental erosion than animals grazing comparable vegetation on less sandy soil or less dusty plants. The cause of this variation is the extra amount of abrasive particles that are introduced into the mouth by the higher sand content of the soil and vegetation.

Notwithstanding this, tooth wear is used almost universally for "age" determination in game management through the development of local time scales for the observed pattern of wear. The basis of these time scales is animals of known age or wear group sequences (Fig. 8) from the area over which it is desirable to apply the scale. The use of tooth sections has, however, shown that some local time scales may be inaccurate, especially with increasing age and this doubt has been reinforced by observation of the wear in samples of known age. Since in management it is often very difficult to apply scientific checks to each animal shot, the animals continue to be aged according to a series of specimen jaws, but it is recognised that although it is convenient to talk in terms of years what is in fact meant is that a given jaw of unknown age belongs to a group showing a certain degree of wear. For examples of this and a clear demonstration of the methods and problems involved the reader should consult the article by Robinette, Jones, Rogers and Gashwiler (1957) and also the paper by Quimby and Gaab (1957) who deal with the problem in regard to elk and mule deer in the Rocky Mountains.

The sequence of wear observed in any population is normally of general applicability (though there can be breed or race differences) but the age to be assigned can be variable. As before we are back to the use of group frequency analysis.

It should, however, be remarked that many of the problems are associated with the need to age precisely animals which may be up to twenty years of age. When attempting to age animals under about five years of age many of these problems disappear and the eruption and early wear of the teeth gives an accurate estimate of age. The mineral intake of an animal has a profound influence on the hardness of teeth and their resistance to wear and decay. Where critical environmental factors such as mineral intake, vegetation and soil are not greatly different in two separate areas there is no reason to expect great differences in the rate of dental attrition in the same

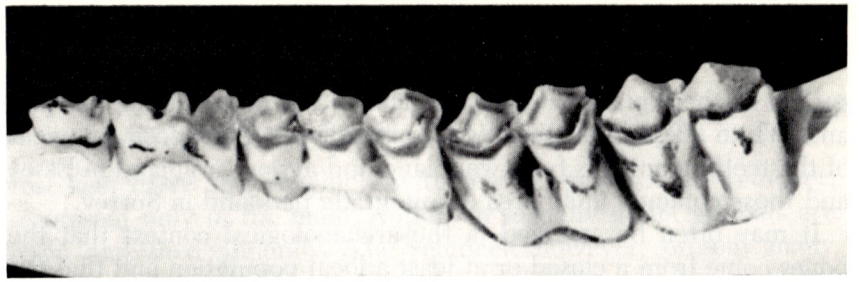

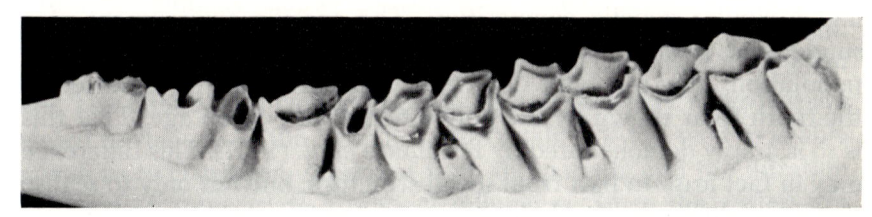

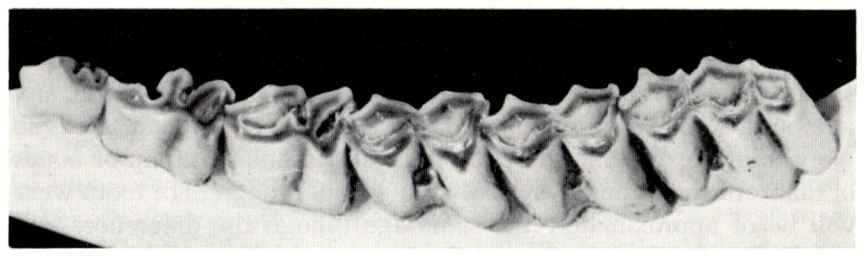

(a)

(b)

(c)

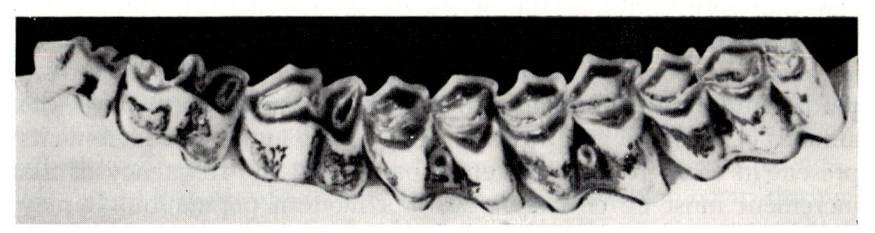

(d)

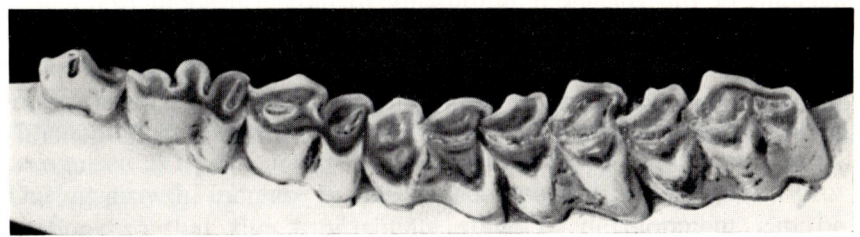

(e)

Figure 8 A series of jaws which form an age index series based on the eruption and wear of the teeth. These are female roe deer all killed between November and February. The first jaw is from an animal in its first year and the others nominally represent successively older animals.

and these can form standard authorities for these species. There is, however, only one major collection of defined measurements relevant to most large mammals, that of Duerst (1926). For many years this has been regarded as a basis for the measurement of bones since it adequately defines many dimensions though makes no attempt to assess their significance. Although Duerst is still an invaluable source, a number of the measurements have been modified by subsequent workers in the light of their experience and accordingly specifications of additional measurements are widely disseminated through the literature. Copies of Duerst are difficult to obtain and a knowledge of German is essential in order to understand the specification of a measure. Clearly another definitive volume on bone measurement is long overdue and progress in animal archaeology will be hampered until this is forthcoming.

It will be appreciated that the publication of measurements is important but space in archaeological journals is at a premium. Accordingly there must be occasions when the specialist must carefully consider the cost of publication of the measurements. It is not the publication of the measurements alone that is the problem but the associated need for space for their definition.

There can be no question of the exclusion of measurements where (a) these can be defined by citation or with a limited amount of extended definition, or (b) when the measurements are of considerable significance because of their character or number, even if detailed definitions are necessary. There are, however, occasions when a number of different dimensions could be measured but there are so few examples of each that their value is limited. It is in cases like this that it may be permissible to omit publication of the measurements since suitable bones have been preserved for future reference. It is very doubtful whether there is much justification for the publication of measurements which are too few for statistical purposes or where the biological significance of the measure is not understood.

It is clearly a matter of some urgency that the principal significant dimensions should be defined and identified by a simple code reference to facilitate publication of the measurements and to ensure that everyone means the same thing when they speak of length or width, which is not the case at present.

Measuring a Bone

Having defined the precise points between which the dimensions are to be determined these now have to be ascertained. The instruments used for measuring will depend on the type of bone and its size. For the larger domestic animals, cats or horses, all that is required are some callipers, dividers, fine tape, scale and a measuring board. For measuring angles a device akin to a clinometer is necessary. The measurement of bones of small mammals like voles and shrews is best done with a travelling microscope or a stage micrometer fitted to a low power microscope and for the measurement of the teeth of such animals they are essential.

Large callipers and those reading to several decimal places are extremely expensive and are quite inappropriate for our type of work. All that is required for general use is a pair of inside/outside engineer's callipers (Fig. 9) with a spread of six to eight inches, with a metric scale and a vernier reading to 0·1 mm. At present this sort of calliper can be purchased from a good tool shop for under two

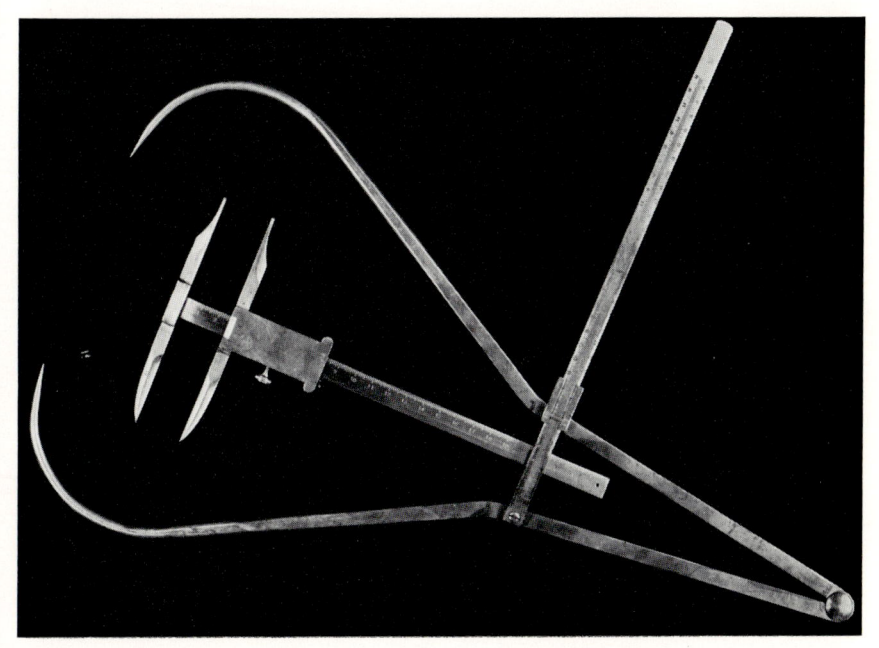

Figure 9 Osteometric equipment. Large external and smaller general purpose callipers.

pounds. A variant of the instrument described fitted with a dial gauge is available and this is certainly easier to read, though I would expect it to be much more expensive. Most of the cheaper callipers are made from mild steel which rusts very easily even indoors let alone in the field. It is therefore well worth paying extra for a pair made of stainless steel. Callipers are not normally used for the measurement of length except perhaps for the measurement of the limb bones of small mammals. They are normally used for relatively short widths or distances which require the use of a more accurate scale or where the shape of the bone is such that the calliper is the most accurate and convenient instrument.

A tape is required for measuring circumferences and also for measurements of length along a curved or irregular surface such as a horn core or rib. The ideal tape should be narrow and soft and graduated in millimeters. Such tapes are very difficult to come by and it is often necessary to make do with a piece of ordinary tape or string and a scale. A disadvantage of the graduated linen tape is that they are liable to warp and wear, especially under difficult circumstances. A metal tape (which should be of stainless steel or other non-corrosive metal) is often the most satisfactory for general purposes. The metal or plastic tapes sold by many shops for handicraft purposes are rarely suitable for measuring bones as their rigidity makes them difficult to adjust to the contour of a bone.

An osteometric board is an essential item but unfortunately these are not easily obtainable. They also tend to be expensive so it is better to make one's own (Fig. 10). The board is simple to operate

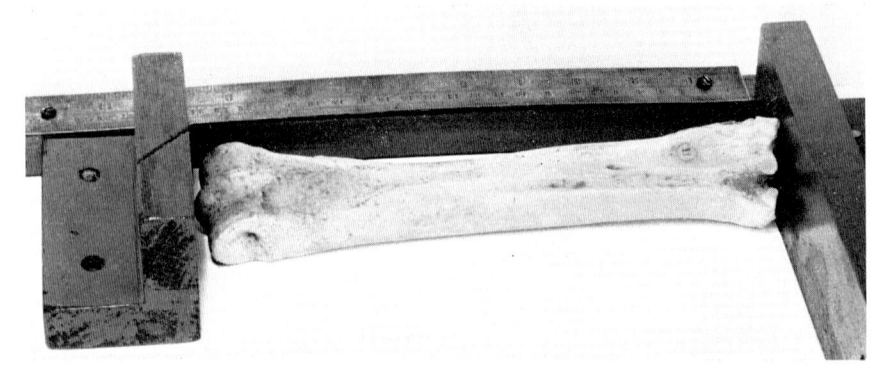

Figure 10 Osteometric equipment. A home made osteometric board as described in the text. Such boards are used for measuring the length of large bones.

and easy to construct. It consists of a permanently mounted scale with a fixed block at right angles to the scale which represents the scale zero. The bone to be measured is orientated on the board with one end against the fixed zero block. A free-travelling block with its face also at right angles to the scale is slid into contact with the free end of the bone and the length is then read off from the scale. An advantage of this type of board is that it permits great freedom of choice in the selection of the orientation to be used.

In constructing such a board the most important feature is the base, which must not warp and which must also be rigid. The most suitable material is probably blockboard which is of composite construction designed not to warp or shrink. Well-seasoned hardwood that will not distort is very difficult to obtain nowadays but occasionally an old piece of furniture may supply such a board. A good quality drawing board would also be very suitable for the base. An inch square strip of wood runs the length of the board on one side and to the top surface of this is attached the scale set fractionally back from the inner edge to avoid contact with the moving block. The scale which I have used is a two foot steel rule with the metric scale in such a position that when attached to the strip it lies along the inner edge. The blocks must also be of quality timber that will not warp or distort, and must of course be accurately machined. An ideal block is a mitre block obtainable from most tool merchants. The fixed block of the board is one of the larger mitre blocks whilst the travelling block is smaller. With the scale and strip firmly fixed in position the block is set up so that its face is at right angles to the zero mark of the scale (not necessarily the end). When set up accurately the block is fixed firmly in place. Glue permits a more accurate positioning of the block than do nails or screws. The small block moves freely against the side strip with its longer machined face at right angles to the scale. The face of this block represents the datum line for reading the scale and this is made more accurate and straightforward if a pointer is added. It was found that a scalpel blade attached to the block face so that the butt end overhung the scale a fraction above it (to prevent it rubbing the scale) made the most satisfactory pointer. Treated carefully a board of this sort will measure accurately within the limits required almost indefinitely.

The principal use of the second type of calliper (Fig. 9) or divider is for the measurment of distances too large or inconvenient for the smaller inside/outside calliper. Large pairs are invariably necessary and if these are not available as a stock item from a manufacturer it may be worth having them made by an engineer. I have found little need for these large callipers with the fragmentary bones from sites, they are, however, of value when recording complete skulls or comparative material.

Various machines, basically frames with travelling points coupled to micrometers, have long been in use for the measurement of human skulls. The machines are, however, not suitable for the measurement of most animal skulls. The majority of measuring instruments needed for the larger animals can all be obtained from engineering sources, and other engineering tools, such as scriber blocks, can often be modified to meet particular needs.

For work on small mammals the use of a microscope is frequently essential. The item may be measured by one of two methods. With a travelling microscope a graticule mounted in the vision tube of the microscope traverses the dimension, the traverse distance being read from a micrometer. Such travelling microscopes are used for accurate measurement in many fields. The second method is that normally used with a low power microscope where the specimen is moved on a mechanical stage across a fixed graticule mounted in the vision tube, the traverse being made and recorded by the stage micrometer.

There is no doubt that far more elaborate measuring devices are technically feasible. There is no reason, for example, why a machine should not be developed to scan a bone electronically or optically and feed dimensions into a computer. By such a method it would be possible to obtain rapidly sufficient measurements to describe the bone to the machine which could thus be programmed to identify it. Such a device could, apart from identification, be a powerful tool in the identification of morphological differences associated with age, sex, breed, etc, and it is with this aspect of recognition that it would have its greatest value.

Measuring Technique and Accuracy

The accuracy of a measurement will vary according to what is being measured and the particular problems involved. The less well-defined the points between which the dimension is to be measured the more difficult it becomes to obtain repeatable results. Thus measurements between points that are ill-defined must have a larger inherent variation than those between unambiguous points. Where the points cannot be precisely defined then any two or more workers are likely to produce different results both individually and collectively. Such variation is no great problem inasmuch as anyone concerned with measurement is aware of its existence. The important point to remember is that even if two people say they are measuring the same thing they are often not doing so precisely. If two sets of figures are found to be different for what purports to be the same dimension it is desirable that one should check to see whether any of this is due to measurement variation and not to real differences between the samples. A valuable discussion of the nature of measurements and their properties is given by Simpson, Roe and Lewontin (1960) and the reader should be familiar with the relevant chapters of this work before undertaking the measurement of bones and other items, or making use of such measurements.

In order to exert some degree of control over the errors inherent in bone measurement—systematic, instrumental and personal—it is necessary to develop a personal technique for measuring bones. My own practice in dealing with a site is to set aside all the measurable bone and then to measure each bone type as a group. Thus all scapulae, tibiae, etc, are measured at one and the same session with the same instruments. If comparative measurements are necessary one endeavours to obtain these at the same session. In the case of adequately defined measurements where the points are unambiguous this is not necessary.

The groups that one has to deal with from archaeological sites are frequently quite small and problems of measurement fatigue are not likely to be encountered. If, however, one is measuring a long series of bones errors associated with eye strain, fatigue, poor lighting, etc, can creep in. Errors owing to fatigue are not usually recognisable unless they are gross ones and thus it is necessary to take precautions against fatigue errors from the outset since it may be very difficult to detect that one's measurements became less precise as the day wore on. The most important point is to establish a technique for measuring that will reduce the errors involved. The way in which the measurements are carried out should be precisely defined in the notebook so that anyone in twenty years' time will know exactly how you measured a particular item which may then no longer be extant.

7
The Determination of Sex from Bones

The sex of an animal can in general terms be identified from certain bones if particular conditions are fulfilled.

1. If unique secondary sexual characteristics are present such as the antlers and bony pedicles of male deer (except in the reindeer, *Rangifer tarandus*, in which both sexes are antlered), the baculum (penis bone) of male carnivores and the metatarsal spurs of some male birds.
2. If there are morphological differences between the sexes as in the form of the pelvis.
3. If the sexes are dissimilar in average size or weight and these differences are reflected in the bone dimensions.
4. If there are qualitative differences in anatomical features such as the lengths of processes and the sizes of muscular ridges between the sexes.

The first two methods are concerned with specific criteria and should be largely unambiguous provided that they have been worked out on sufficiently large samples for the species involved. The third and fourth methods are concerned only with differences of degree and are not absolute, therefore some degree of overlap in the range of values may be expected. Taking the dimensions alone it may be found that only extremes are separable with certainty. Further, as a result of environmental factors, a dimension characteristic of the males of one population may be a property of the females of another. The application of methods of sex estimation to bones from excavations presupposes a knowledge of the variation in a known modern population of that species, or alternatively upon a subjective interpretation of the variation observed in the archaeological population. It is, of course, always questionable whether the results of studies on modern domestic stock are applicable to ancient stock. This doubt is not only occasioned by the selective breeding of modern stock, but even in unspecialised breeds by the separation in time of the two populations.

The use of crude dimensions to separate the sexes of different populations by size has many points of objection. In domestic stock, given the size variation found between present breeds, it would clearly be absurd to set dimensional limits to the size of bone for each sex. The same objection would also apply to wild populations though in practice it may not always apply. In the case of fallow deer metacarpals from different populations in the British Isles, size would appear to be an efficient criterion of sex.

To overcome these objections many authors have attempted to establish indices which would separate the sexes and be independent of variations in the size of the animals. In many cases the separation is not complete and there is an overlap in the values for the sexes. In domestic stock the picture is further complicated by the presence of three sex categories—intact males and females and castrated males. For castrated male cattle the average values of metacarpal indices are intermediate between the values for males and females and overlap into the range of each. A further difficulty is encountered in interpreting indices from these animals since the relative position of the castrate between males and females will be influenced by the age at which the animal was castrated.

In attempting to sex the bones from excavations it will be found in practice that methods one and two can rarely be used and for an analysis of the sex structure of the population recourse must be had to either method three or four. Using method three it is at least possible to present the facts in a precise manner, whilst method four is entirely subjective and offers the reader of the report no opportunity to assess the validity of the sex estimates. For this reason I prefer not to make general use of method four when reporting this material.

On most excavations only a small proportion of the bone is intact and thus the use of indices for sex estimation of the whole sample is often impracticable. Recourse must invariably be made to the statistical distribution of particular bone dimensions. A common method of presenting these data is the histogram. Figures 12–14 (p. 106) show typical results in which a number of groupings of the measure will be observed. By the application of statistical tests it can be shown whether these groups are significantly different or are but part of a continuous series from which examples are missing. In applying these tests it should be remembered that in fact the whole sample is one population; it is the individual groups which should be tested to see whether a subdivision into females, castrates and males can be upheld. There can, however, be no rules for the interpretation of these histograms since in some cases it may be possible to sex a number of bones by the use of an index and thus include in the histogram bones of reliably estimated sex. A word of caution is, however, necessary. Although the histograms may show distinct groups it does not follow that reading from left to right, these correspond to females, castrates and males. It could be the reverse order. Accordingly it is recommended that the worker should establish the character of the variation and its direction by repeating the measurements on the bones of modern animals of known sex.

In interpreting the histogram a negative separation is as significant as a positive one provided that the measure can be shown to have a discriminatory value by reference to modern animals. It is in fact extremely likely that sites with a particular economy will tend to produce bone refuse from animals of one sex. This would be the case with sheep where wool and milk were the prime concern and where surplus males were perhaps marketed on the hoof. This situation is in part suggested by the study of the Saxon bones from Whitehall, London, where other evidence supports this interpretation of the graphs (see Chapter 9). Comparatively few studies have been published on sexual dimorphism in bones of modern species. With the larger animals, in particular the ungulates, much of this published work has been concerned with the metapodial bones. This bias towards the use of these particular bones in investigations of modern species is not necessarily a reflection of their particular suitability for sex determination but rather a matter of economics. The metapodials are the bones which are discarded when a carcass is dressed whilst the other limb bones are sold as an integral part of it and are thus much more expensive to obtain from modern animals. From the archaeological aspect it is the metapodials which probably survive intact far more frequently than any of the other limb bones. This of course is also related to the disposal of carcasses.

Zalkin (1960) has studied measurements of the metapodials in the Kalmyk breed of cattle. The sample examined comprised 59 cows, 10 bulls and 13 oxen. He found that although the bulls were taller than the cows the metapodials were of similar length. The metapodials of the bulls are, however, more massive than those of the cows. Sexual dimorphism between the bulls and the cows was more marked in the metacarpal than in the metatarsal. The ratio of the width of the distal epiphysis to the total length of the metacarpal places the oxen between the cows and bulls. The metatarsal has a very limited discriminatory value owing to the large range observed in absolute size and with the indices. (The original article is in Russian and the summary given above is from the short English summary.)

Howard (1963) has examined the differences in the metapodials in a series of modern cattle obtained from slaughterhouses, as well as cattle from archaeological sites. She found that sex related differences were observable in both the modern and archaeological samples. Two indices were used: the metapodial indices DB/L and MB/L.

In the *metapodial index* DB/L, DB is the greatest breadth across the distal joint surface, and L, is the greatest overall length. This is calculated as

$$\frac{DB}{L} \times 100 = DB/L \text{ index}$$

With the *metapodial index* MB/L a third measurement (MB) the transverse breadth at the mid point of the overall length is used in the formula

$$\frac{MB}{L} \times 100 = MB/L \text{ index}$$

From a sample of 66 metacarpals the following range of indices was found. The figure in brackets is the number of examples of each sex.

DB/L Metacarpal
   females   24·8—33·6 (40)
   oxen (castrated males) 29·3—32·9 (8)
   males   32·5—37·8 (18)

MB/L Metacarpal
females 12·9—19·5 (40)
oxen   14·0—18·3 (8)
males   18·6—24·5 (18)

From a sample of 70 metatarsals the following range of indices was found:

DB/L Metatarsal
females 22·1—28·6 (44)
oxen   24·6—27·5 (8)
males   24·8—30·3 (18)

MB/L Metatarsal
females 11·5—14·7 (44)
oxen   12·7—15·8 (8)
males   14·7—19·2 (18)

From Howard's work it is clear that in cattle these metapodial indices have a limited discriminatory value.

Skinner and Kaisen (1947) applied the MB/L index to the metapodials of the North American Bison (*Bison bison*) and found that this index gave a clear separation of males and females. Here, of course, the problem of castrated animals did not arise. Examination of Howard's results shows that the male and female indices alone overlap but slightly. The difficulty occurs in attempting to distinguish the castrated animals.

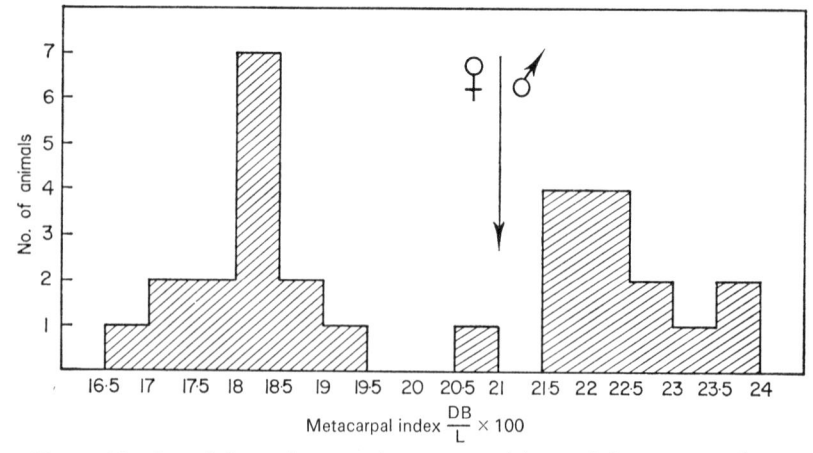

Figure 11 Sexual dimorphism in the metacarpal bone of the European bison.

Empel and Roskosz (1963) give dimensions of the metapodials and other bones of European Bison (*Bison bonasus*) of known sex and age. These figures are plotted in Fig. 11 for all animals over four years of age. It will be seen that the DB/L index produces a clear separation of the two sexes in adult animals; only one animal, an eleven year old female, closely approaches the male index.

Higham (1969) has examined bovine metapodials and some other limb bones to assess their value in sex discrimination. This study is of particular value as it deals with castrated males and females. The most efficient and useful discriminant of sex was found to be the index, *maximum distal width of diaphysis*:*maximum width of distal epiphysis*. These two dimensions are usefully shown in graphic form as ellipses representing the limits of statistical confidence levels for the separation, can be drawn in. The writer has used this index to study the metapodials from a number of sites. Few of the animals studied approached the size of Higham's modern animals but it was found that the dimensions of these smaller animals were distributed in groups on axes very close to those along which Higham's samples were distributed. This suggests that this index is independent of absolute size and is thus a valuable sex discriminant in both castrated and intact animals.

In sheep, Zalkin (1961) has examined the metapodials of 50 wild *Ovis ammon* and *Ovis canadensis*, and 134 domestic sheep of eleven breeds. It was found that sexual dimorphism was weak though on average the metapodials of males were somewhat longer and more massive. The effects of castration were not found. Individual variation, and variation between breeds was considerable and the same dimensions appeared in very different breeds.

From these papers it would appear that in some species the sex of the individual can be estimated from indices or dimensions, but that in others it may not be so apparent. It should, however, be considered particularly with domestic stock, that the results from modern animals may not necessarily apply to the ancient stock. Thus a histogram of dimensions or indices obtained from an adequate sample of modern stock of known sex which fails to discriminate between the sexes need not invalidate the interpretation of a comparable diagram from an ancient population which produces a significant grouping. It may be a character of the ancient population that sexual dimorphism is detectable in the bones. On the other hand the rather irregular groupings often produced by ancient population samples should be interpreted bearing in mind the pattern produced by the same data from a modern population of known characteristics. Howard's figures show, for example, that considerable doubt can be raised as to the attribution of many specimens in a sample that comprises the three sex groups where her indices are used. If we accept that this doubt exists it will necessarily limit the interpretations that can be made. The study of the distribution of a measure is, however, extremely important for it enables a number of probabilities to be recognised. From a study of previous figures it is clear that a bimodal frequency of the sort seen in Fig. 13 suggests two components probably of a single popu-

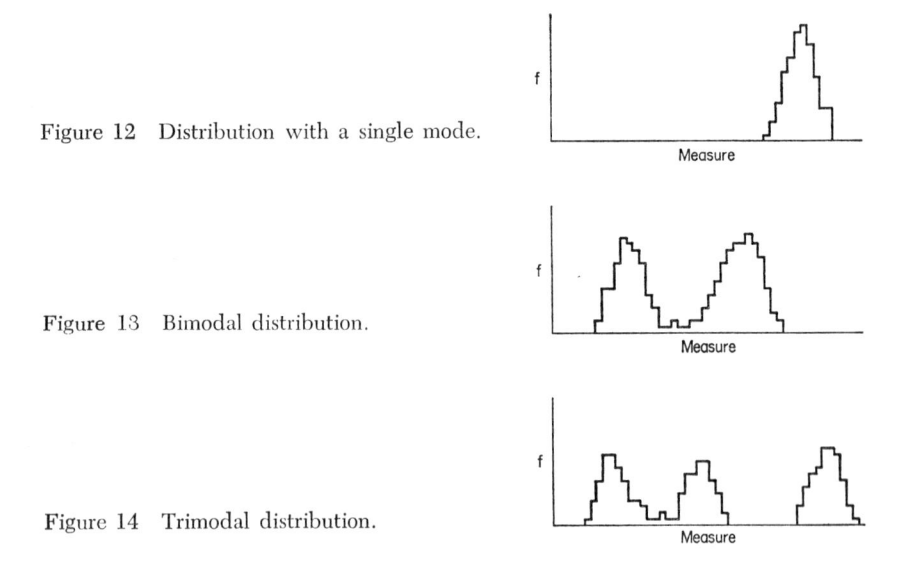

Figure 12   Distribution with a single mode.

Figure 13   Bimodal distribution.

Figure 14   Trimodal distribution.

lation. These could reasonably be interpreted in the case of bovine metapodials as animals from two of the three sex groups, most probably male and female. The distribution seen in Fig. 12 suggests a homogeneous sample probably consisting of only one sex. On the other hand the distribution in Fig. 14 is more likely to be from a population composed of all three sex classes. While some valuable inferences can be drawn from these dispersals or groupings on sites of all periods we clearly expect all sex groups to be present unless our sample is biased by, for example, the use of only castrates for meat. Such a practice seems inappropriate except in an advanced husbandry and marketing economy. If, however, we are concerned with sites of incipient domestication, the mere presence of a distribution as seen in Fig. 14 preceded in time by either of the distributions seen in the previous figures, must argue strongly for a fundamental change in animal exploitation and possibly a transition to domestication. Thus although our figures for the identification of the sex group are open to doubt for many of the individual animals there is a forceful argument to support the subjective interpretation of the presence of males, females and castrates in the sample.

8
Bone Pathology

Diseased bones may be found on archaeological sites but are usually few in number. Occasionally such bones are frequent and clearly of significance in the study of the population. It is therefore necessary that the archaeologist and specialist should be able to recognise pathological lesions in bones and have some knowledge of how these can arise.

The range of bony lesions that can be precisely identified in veterinary practice is much wider than can be determined from archaeological material. This is because both infective organisms and the degree of soft tissue involvement can be determined at post mortem. In archaeological bone we can consider only the lesions of the mineral skeleton. Because different diseases can produce superficially similar lesions in bone a classification of diseases according to the character of the lesion is not practical. The archaeologist and specialist are primarily concerned with the recognition of a lesion, its probable origin, and the information that it can give about the condition of the animal once the type of lesion has been identified by a pathologist. The classification of the diseases presented here follows a causal pattern which is largely coincident with the type of lesion produced. The classification and description of the lesions follow that of Jubb and Kennedy (1963). I have, however, simplified and modified their excellent account of skeletal diseases and disorders according to what I think is applicable to material from archaeological sites. The reader concerned with pathological material and its interpretation is advised to consult this work and the references therein.

The diagnosis of skeletal disease is a difficult matter, and requires appropriate training and skills. It will be appreciated in the account below that similar lesions may be caused by a variety of diseases and several may be active at the same site. A final diagnosis should therefore only be made by a pathologist or person familiar with bone lesions. The descriptions and examples below are not guides to identification but a classification of the diseases and disorders affecting bones and the lesions they produce. The following categories of bone disease and deformation are distinguished:

1. abnormalities of development.
2. metabolic disorders.
3. hormonal disturbances.
4. necrosis and inflammation.
5. neoplastic and similar conditions.
6. bone discontinuities.
7. diseases of the joints.

Abnormalities of Development

It will be appreciated that the complex processes of bone formation provide many opportunities for developmental error. These may affect the whole system or be confined to specific bones or parts.

One of the best known generalised disturbances is *Chondrodystrophia foetalis* which produces disproportionate dwarfism. The defect lies in the interstitial growth of cartilage. Thus longitudinal growth in the epiphyseal, articular and basocranial cartilages is prevented, whilst the appositional transverse growth is normal; hence only bones with cartilaginous precursors are affected. The best known example of this condition occurs in cattle of the Dexter breed in which it is transmitted as a simple dominant autosomal character. Various forms of the condition are known.

Dyschondroplasia is characterised by the development of numerous bony protuberances adjacent to epiphyseal and functionally related cartilages. This is a result of the ossification of excessive lateral proliferations of the cartilage. In mature bones these exostoses should cease development but continued growth suggests that these exostoses have undergone neoplastic transformation. The condition is rare.

Localised disturbances are numerous and in some cases are part of the characteristics of a breed. The short limbs of Dachshund and Basset-hound dogs are due to developmental disturbance of the lower limbs. Other limb disturbances include both partial and total absence or reduction of the limbs, or part of the limb. *Syndactylia* (fusion of the toes) occurs in cattle and pigs and *Polydactylia* (increase in the number of toes) is found in most domestic species and is best known in cats and dogs where it is often a "familial" characteristic.

These disturbances are genetic in origin but deformities also occur which are adaptational in character and by implication appear after birth. These adaptational deformations are a result of the degree of self determination of form inherent in bones. Thus, abnormal loading or displacement of a bone during its formation will result in a shape related to these forces and not to the normal condition of the bone. In mature bone the same processes result in remodelling. Where a bone/muscle system is inactivated this leads to osteoporosis and a resorption. The most frequent cause of this is paralysis following nerve section or demyelinating disease.

Metabolic Disorders of Bone

The formation of bone involves two critical stages which can be affected by metabolic inadequacy or failure. Inadequate formation of the matrix results in only a slight bone formation which is, however, adequately mineralised. This failure is termed *Osteoporosis*. The other condition is the failure or inadequacy of the mineralisation of newly formed organic matrix. This is termed *Osteomalacia*. Osteomalacia and osteoporosis may be present in the same animal and this may be complicated by a third condition, *Osteodystrophia fibrosa*.

The causes of osteoporosis are non-specific and are to be regarded as a consequence of the disturbance of tissue metabolism rather than of calcium and phosphorus metabolism. The normal resorptive processes continue but are not balanced by deposition of new bone. In consequence the bones become porous and light and fracture easily. The medulla enlarges and may extend into the epiphysis. The disease is systemic and generally slow and is not usually observed except as a consequence of prolonged malnutrition or old age.

Osteomalacia is the result of a continued imbalance of calcium (and Vitamin D) or phosphorus. It is a disease of adult bones and the condition in young animals is termed *Rickets*. Because of the differing ages of maturation of the bones osteomalacia and rickets can co-exist in the same animal. Osteomalacia is characterised by softening of the bones due to the resorbed mineralised tissue being replaced by inadequately calcified tissue. The bones break easily and many deformities may arise as a consequence of mechanical failure. The deformations of osteomalacia are less severe than in rickets as the turnover of bone tissue is concerned with repair and remodelling which is less extensive than the tissue turnover associated with growth.

Rickets is an osteomalacia in bone that is still growing. The lesions are most prominent in areas where the cartilage is involved in rapid growth. Thus the shafts of the long bones are shorter and broader than in normal animals and have a narrower marrow cavity. The cortex of the bone is soft so that many deformations involve curvature due to failure of mechanical function. There is enlargement of the ends of the long bones and there may be lipping over of the shaft and epiphysis. The teeth and jaws are also involved and this may give rise to mal-occlusion, abscess formation and excessive and irregular attrition. When the nutritive balance is restored the lesions heal rapidly and some remodelling of the deformities occurs. The majority of the deformations are permanent, however, and will often give rise in later life to other conditions such as joint disease.

Osteodystrophia fibrosa may be superimposed on either rickets or osteomalacia as a nutritional disease. It is marked by rapid osteoclastic activity with exuberant growth of fibrous tissue in the bones. It can be provoked by diets deficient in calcium or with an excess of phosphorus. It can, however, arise from *hyperparathyroidism*. The precise circumstances in which the condition occurs in association with rickets or osteomalacia are uncertain.

These disorders which may be related to conditions in which the animals are kept are of great interest in the archaeological context because of their association with husbandry technique and the availability of foodstuffs. Careful examination should be made of bones for evidence of these conditions. As these disorders vary in severity and the original lesions can in part be repaired the extant scar may be slight. These lesions should be interpreted by reference to the causal diets and conditions, compared with the food plants available, and any structural evidence of byres, etc.

Osteodystrophies are also associated with other causes such as primary hyperparathyroidism, renal insufficiency, Vitamin D poisoning and fluorine poisoning, whilst others are of unknown origin.

Hormonal Influences

The effect of the parathyroid has already been mentioned. Inadequate supplies of growth hormone from the pituitary produce a dwarfism which must be distinguished from that of genetic origin. Hypopituitary dwarfs have delayed bone development and epiphyseal closure. Hypersecretion of pituitary growth hormone produces gigantism which again must be distinguished from that of genetic origin.

The thyroid gland also influences development and in *hypothyroidism* there is retardation of growth rate and proportional changes in cartilage bones. The hypothyroid dwarf has infantile skeletal proportions, delayed epiphyseal fusion and epiphyseal dysgenesis. *Hyperthyroidism* produces an acceleration of growth and maturation in the juvenile and is often associated with osteoporosis probably the result of a metabolic imbalance.

Both oestrogenic and androgenic hormones produced by the gonads influence epiphyseal closure and bone development. They are responsible for sexual dimorphism in form and size. Thus castration or gonadal disease will affect the normal sexual differentiation of the skeleton and the development of the secondary sexual characteristics. The effects vary between species. In general castration of the young male leads to greater length of long bones since epiphyseal closure is delayed. The skeleton is less rugged than that of the normal male and is difficult to sex.

Necrosis and Inflammation of Bones

Necrosis of bone (*osteosis*) frequently accompanies injury in which the blood supply to an area is reduced. The diagnosis of dead areas in a fossil bone will depend on the extent to which the dead area has been repaired. Osteosis is usually initiated by fractures which disrupt the periosteum, injury to soft tissues and infection. Healing consists of the removal of dead bone and its replacement by new bone tissue. Erosion and deposition proceed together so that at any one time there may be quite a large swelling of fibrous new bone which in time will be remodelled. Where damage is extensive the new tissue may completely enclose the old and resemble a tumour. Necrosis may also occur from non-traumatic lesions; these include ergotism, fescue foot and chronic anaemia.

Osteitis refers to inflammatory changes established within bones, An osteitis in which the seat of the infection is the periosteum is termed an *osteoperiostitis* and one in the medulla an *osteomyelitis*. Both may develop and extend into adjacent regions.

There are many organisms which produce osteitis either by infection through the bloodstream, from surface injuries or from lesions in adjacent soft tissue. Osteitis is common in many animals at the present and the lesions may be localised or extensive. Some have little effect on the animal whilst others may cause considerable damage and marked impairment of feeding ability. A typical causal organism of mandibular infection (lumpy jaw) in cattle is *Actinomyces bovis*. This infection is termed *Actinomycosis*. A variety of names are applied to these conditions according to the organism responsible. The clinical lesions are, however, broadly similar. The diagnosis and naming of the condition depends on the identification by appropriate techniques of the organism involved. As the causal organism cannot be detected in archaeological material diagnosis of such an osteitis is confined to gross description of the lesion and the progress of the infection. It is therefore inappropriate here to consider further the range of organisms associated with osteitis in various species.

Neoplastic and similar conditions

Both benign and malignant tumours occur in and around bone tissue. Their development provokes the usual osteoclastic and osteogenic response but the appearance of the tumour is highly variable depending on the tissue involved and the progress of the disease. Those developing within the bone may not be visible externally and like cyst formations can only be detected by X-ray photography.

Hypertrophic osteopathy is characterised by the formation of a diffuse osteophyte deposit over the periosteal and to a lesser extent the endosteal surfaces. The pathogenesis of the condition is unknown but it is frequently associated with chronic pulmonary disease. A localised form of this disease is found in the cranium and mandible of some breeds of terrier dog, e.g. the West Highland. Here it is characterised by extreme bony proliferation which stops at the end of normal bone growth. The cause of the general condition is unknown.

Bone Discontinuities

Traumatic fractures occur as the result of injury to healthy bone. Where, however, the bone is diseased pathological fractures can arise spontaneously. The healing of traumatic fractures involves the formation of new supporting tissue and the reorganisation and repair of that damaged. A treated fracture can heal readily and in time may leave little external indication of the event. When, however, the fracture is severe, involving the displacement of the parts and spontaneous repair, the proliferation of new tissue may leave a gross deformation. Where there is severe damage to adjacent soft tissue infection frequently sets in. Here reparation is impaired by the course of the infection and the structure of the affected area is complex. Pathological fractures rarely heal until the causal imbalance is removed.

Diseases of the Joints

Congenital hip dysplasia is an hereditary condition affecting the hip joint and is associated with architectural defects of the acetabulum (socket) of the pelvis and the head of the femur. The degree of deformity varies, but in severe cases the acetabulum is shallow and the head of the femur may be grossly deformed. Secondary infection can occur resulting in various degrees of exostosis.

Degenerative arthropathy is a function of the ageing of joints and is to be distinguished from changes associated with arthritis which are of infective origin. The primary failure lies within the cartilage of the joint and the condition is only recognised archaeologically at the next stage when secondary changes have occurred in the bone. This results in localised proliferations of bony tissue which are kept smooth by joint action. Proliferative lesions may also occur at the periarticular sites giving rise to "lipping" at the edge of the joint surface. The interarticular lesions appear slight in character and close scrutiny is necessary for their detection.

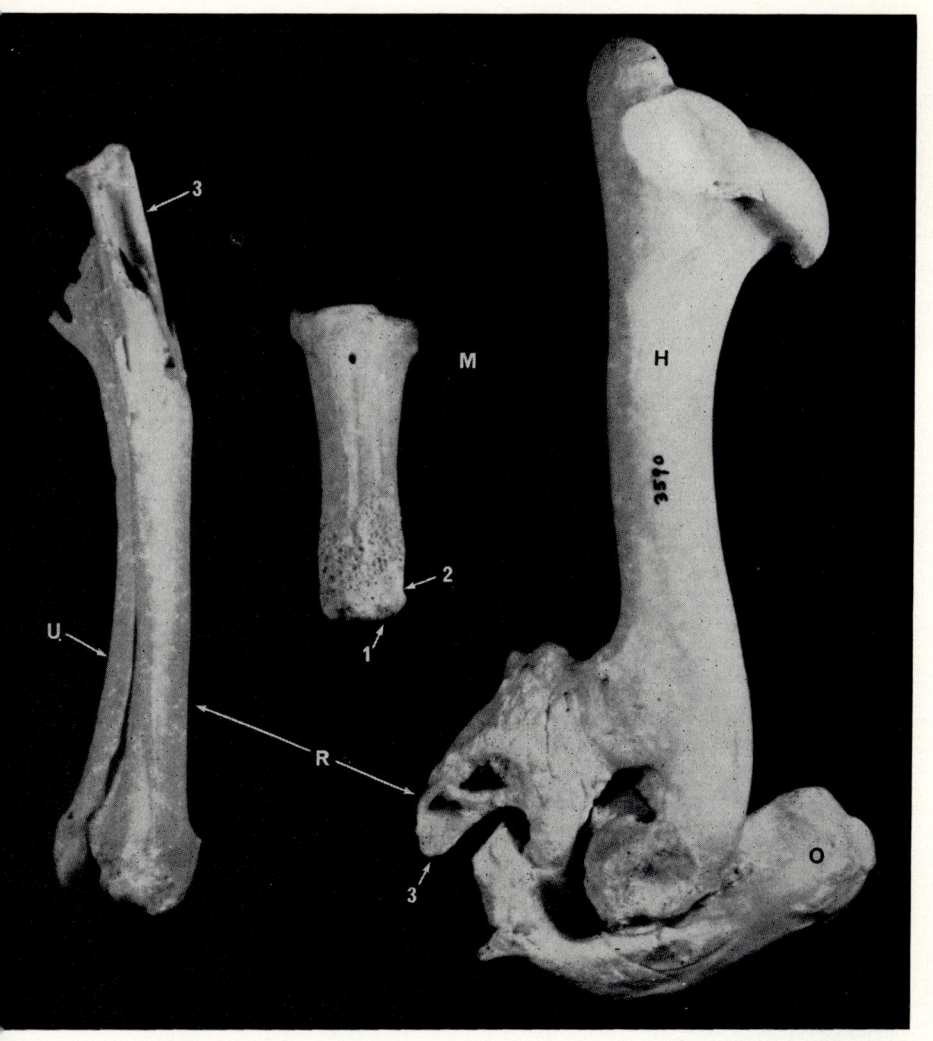

Figure 15 Multiple lesions which are healing spontaneously. The forelimb shown here is from a fallow deer which was killed in a road accident. The lesions seen are the result of a previous injury but the fractures were sustained in the fatal accident. The limb was inoperative and folded under the animal. In the previous accident the elbow joint was extensively damaged and the distal end of the humerus and the proximal end of the radius and ulna were each broken in several places. The lower end of the metacarpal was also shattered. This injury repaired in the wild as follows. The limb became inoperative and the lower leg from one third of the way down the metacarpal was lost. The stump of the metacarpal has healed over (1) but is being resorbed (2) as it is no longer functional. The lower shaft of the radius and ulna is unaffected. The shattering of the elbow joint has resulted in all three component bones being immobilised and united by a proliferation of bony tissue. All have united in a displaced position. In the fatal accident the radius has broken at the point of fusion (3). This limb would never have become functional.

h=humerus; o=olecranon process of the ulna; r=radius; m=metacarpal; u=ulna. A less damaged example of this joint is shown in Figure 16.

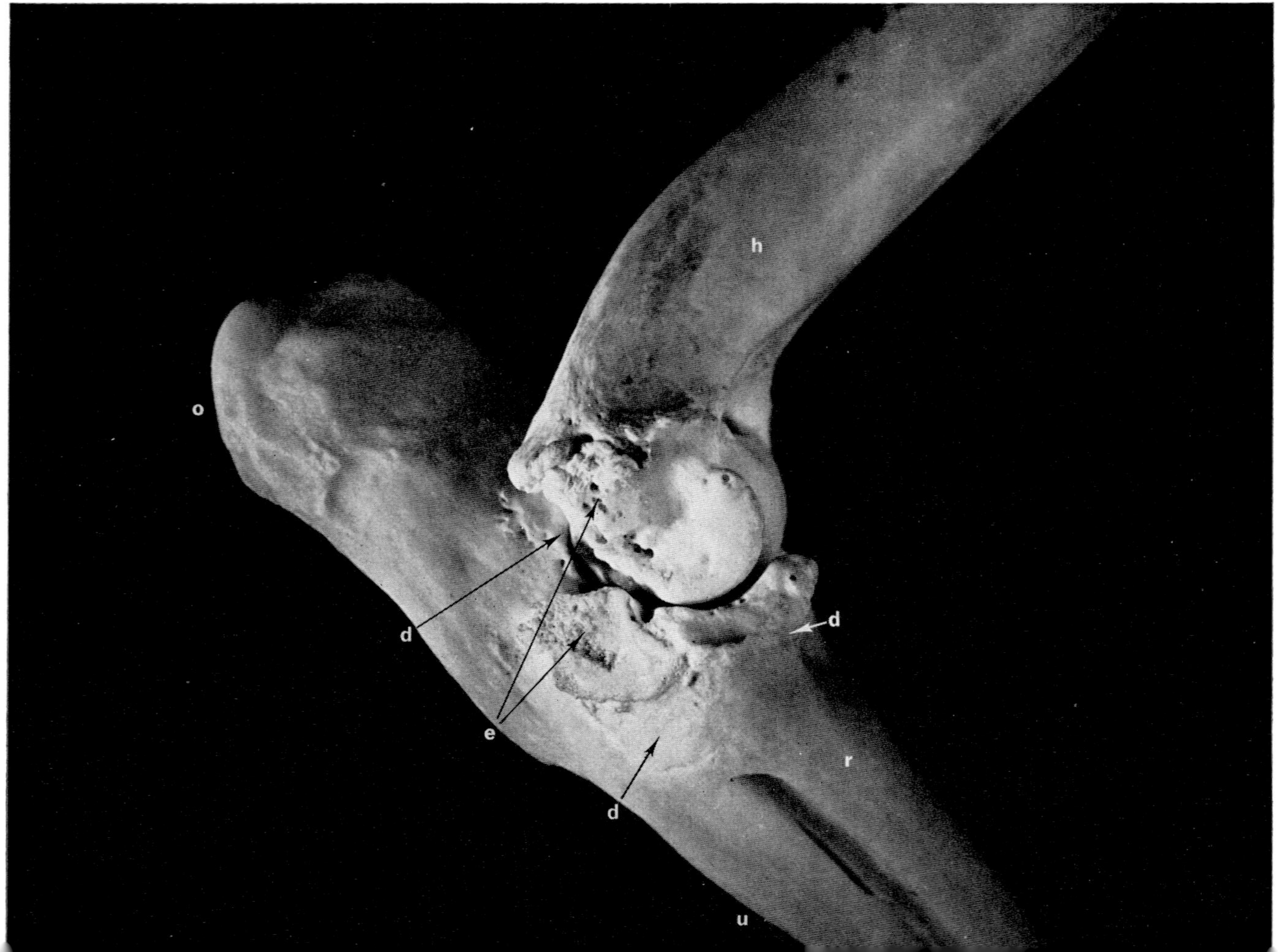
h
o
d
d
e
r
d
u

Ringbone is a specific degenerative arthropathy of the phalangeal joints. The name is derived from the marginal exostoses which are characteristic of the condition. Where two bones are involved they may become united at the margins, thus immobilising the joint. The condition arises from repeated minor mechanical stresses and is characteristically found in the forefeet of horses.

Spavin is an arthropathy of the hock joint of horses and cattle which may result in fusion of the joint.

Infective arthritis is by definition restricted to the interarticular surfaces of the bone, i.e. those enclosed by the joint capsule. Once established a suppurative infection causes severe damage to the articular cartilage and subsequently spreads into the bone and throughout the capsule. The condition is characterised by erosion of the articular surfaces and as the infection spreads may also include the joint margin and adjacent areas. Both erosion and proliferation of bone occurs and the joints may become partly fused. The mode of entry of the infective organism makes little difference to the gross appearance of the lesions. Arthritic changes may be superimposed on previous injuries and infections.

These disease categories are applicable to all mammals but the incidence of a disease may vary between species, populations and with time. In a wild population traumatic lesions are likely to increase when it is exposed to human predation. Equally, the keeping of livestock and pets could increase the frequency of metabolic disorders. For many years zoo carnivores fed on boneless meat suffered from bone diseases due to the imbalance of calcium and phosphorus in their diet.

Little is known about the incidence and distribution of disease in ancient animals. There is much that could be learnt from a survey of pathological conditions in archaeological bone. In particular one notes the relationship between metabolic disorders and husbandry practices, the traumatic lesions associated with activity (e.g. ring-

---

Figure 16 Periarticular lesion of the elbow joint of a fallow deer. This joint was fully mobile and the lesions are confined to the margin of the joint. The lesions comprise deposition (d) and erosion (e) of the bones forming the joint. The ulna is firmly united with the radius. Such lesions are often seen in archaeological material and at this stage probably have little clinical effect. In time however such infections may spread and affect the articular surfaces causing considerable discomfort. This is the same (elbow) joint as is shown in Figure 15.

o=olecranon process of ulna; u=ulna; h=humerus; r=radius.

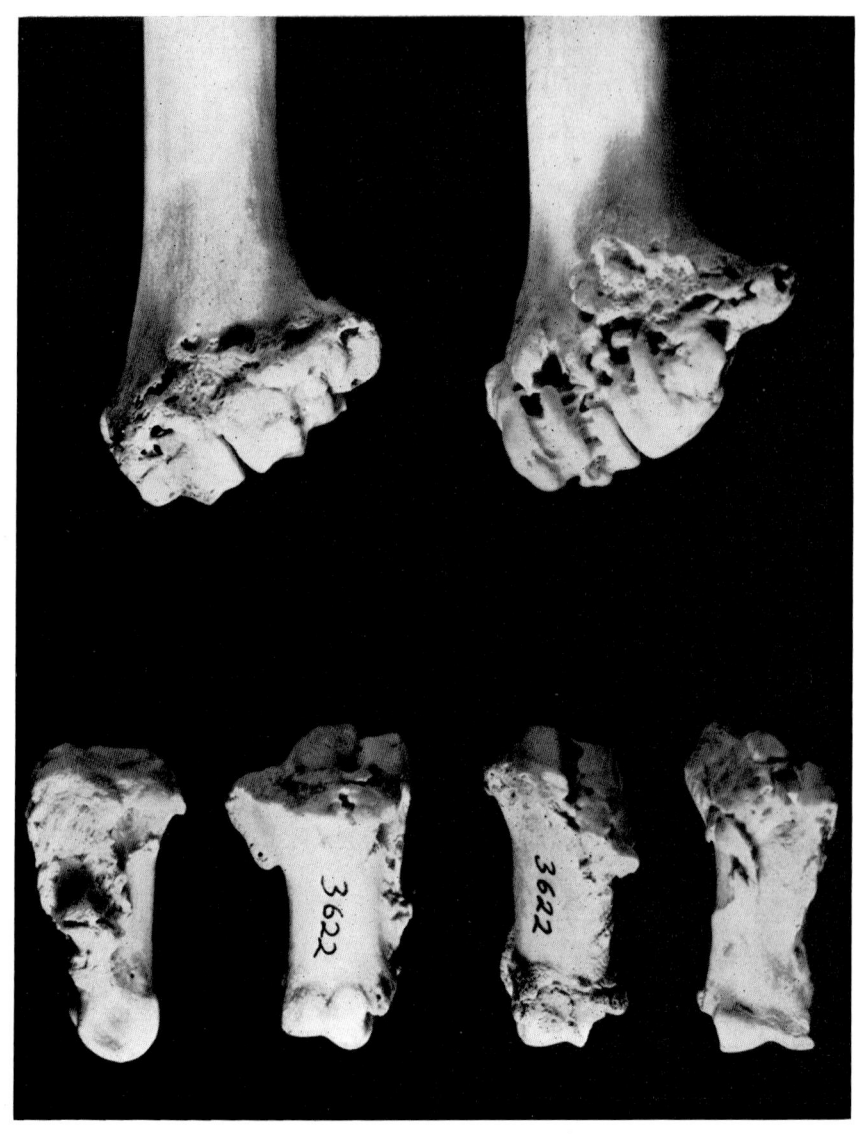

Figure 17 Extensive lesions of the lower cannon bone and first phalanges of a fallow deer. This particular deformation has been found in several fallow deer from different parts of the country but its cause is not known. The lower end of the cannon bone (metacarpal) shows extensive erosion of bone which has led to the collapse of the epiphysis and subsequent proliferation of bone. Areas of erosion and deposition are seen on all surfaces of the first phalanges and are clearly associated with the joint.

Figure 18 Lumbar vertebra of a fallow deer. The left transverse process has been broken (1) and has subsequently healed with some displacement of the process. The articular surface of the centrum shows extensive erosion (2) and there is a proliferation of bone at the joint margin (3). This probably arose as a result of injury and subsequent infection of the joint capsule. Similar conditions are noted in archaeological material especially with horses. Its effect would depend on the progress of the infection and the number of other joints affected as well as the use made of the animal.

bone in horses) and the local proliferative or resorptive lesions associated with pressure such as might be caused by a tie rope. On a wider front it would appear from an examination of archaeological material that lesions are fewer than in some modern populations. Chronological study of the incidence and associated environments of particular lesions might indicate why this is so.

It is emphasised again that the diagnosis on which such surveys are to be based must be made by an appropriate expert in the same way as the identification of the environment of the ancient animal is made by the excavator and his associated specialists.

9
The Interpretation of Bone Evidence

The Limiting Factors of the Archaeological Evidence

The remaining chapters of this book are concerned with the interpretation of the information derived from the examination of the bones and from the ancillary evidence that has been collected. Before we can interpret these data we must understand two things about our collection. Firstly, we need to know the origin and nature of the bones and, secondly, where these are refuse how representative this is of the rubbish from the site. Clearly much of this is inference and we therefore have only a weak foundation on which to erect our intrepretation.

Refuse deposits found on sites might be grouped as follows: closed structures such as pits or wells, open areas such as middens, derived refuse and refuse accumulations over floors, etc. The first of these present few problems of interpretation, they are clearly a deposit with a limited date range, the true time range of the final accumulation probably being far less than that implied by the archaeological dating limits. They are, of course, terminal accumulations for we do not know how often they were cleared out and the muck deposited on the fields. Middens tend to accumulate over long periods and only scrupulous attention to the stratification if any can yield a satisfactory picture.

Derived material or rubbish survival is the bane of the excavator and specialist alike. It is normally only detectable by reference to pottery or other datable objects. Where contamination is so detected a date cannot be assigned to individual bones and therefore these contaminated deposits must be ignored when seeking economic

9. THE INTERPRETATION OF BONE EVIDENCE

evidence, etc. Depending on the date range of the contamination such deposits may still be used for zoological work if worthwhile specimens are available.

The refuse accumulation over floors is clearly a terminal one and it is very debatable whether this represents squatter occupation or merely a failure to clear up the mess before departure. The latter alternative presupposes that the inhabitants led a particularly squalid existence, since the surviving depth of refuse indicates a much greater original thickness. The wide range of sites on which such accumulations are found, the thickness of the original deposit and the accompanying stench, leads one to suspect that these deposits are not associated with the last period of occupation but may represent the use of the area as a deliberate refuse tip.

Many would disagree with this view of the origin of such accumulations and their relationship to the occupance. The problems of refuse accumulation and disposal are too important in archaeology to be left like this and the time has surely come to apply some more conclusive tests to the problem. Comparisons might, for example, be made with sites that came to a catastrophic end and those that just fell into disrepair. It is time that hypotheses were replaced by facts.

The second problem is how representative our collection of bones is of the total refuse? The answer to this must in part lie with the extent of the excavation. Given the areal extent of the excavation we must bear in mind the probable dispersal of refuse from the site. On a cost of time basis we must surely accept that even if refuse was strewn over the floor it would have been raked out occasionally. At one level it probably went little further than the door. At a higher level it no doubt went to more specific areas, to fill in irregularities of the site, old pits, disused wells, etc, into specially dug refuse pits or on to a midden. Some of these would remain undisturbed, others might be cleared out. Such clearances, and also the use of a midden probably mean that some was carted away and presumably spread over the fields as manure. We are unlikely to obtain much of the dispersed rubbish and will clearly never know its quantity. What we can often be reasonably certain about is that the material we do recover is a fair cross section of what was dispersed. We do not know the proportion but we can make a reasonable estimate of its nature. This is a fundamental limitation to our work.

It does not, of course, follow that if, say, a third of the site has been excavated we have a third of the refuse from the vicinity. This is because in any, even a crudely organised disposal system, there will inevitably be areas of excess accumulation. It is very necessary that the archaeologists should not be satisfied with an excavation only of the main structure but that he should pursue his investigations well into the backyards where refuse may more reasonably be expected to have found its way. Where refuse deposits are encountered it is important that they should be stripped out in their totality so that as much as possible is available for examination.

"Workshop" accumulations of bones are much easier to interpret (see, for example, Chaplin, 1966). The same principle applies here that every effort should be made to recover the total volume of debris.

It is of interest to refer here to the beginnings of an extremely important long-term study of the food and debris of a Roman town. At Wroxeter, Dr Webster has found in the forum area a swimming pool that was never finished but was filled in with refuse. There is a strong implication here that a municipal refuse disposal system was in operation. A detailed study of the contents of this town dump is being made and it will be interesting to see in years to come the nature of any contemporary refuse accumulations associated with individual habitations in the town. If the interpretation of this as a town garbage dump is correct it offers tremendous opportunities for a study of town and country economics.

The Use of Ancillary Evidence

The wider interpretation of the bone evidence depends both on the knowledge and experience of the investigator and the extent to which it can be integrated with evidence from other fields. The information obtained about the plant species present and the habitat they formed may, for example, be fundamental to any understanding of the husbandry. The identification of the species utilised is not always dependent on the bones. In some areas, e.g. south western U.S.A., feathers may be preserved (Hargrave, 1965; Messinger, 1965) and the evidence from both sources must be considered.

A study of the history of livestock characteristics must take into account the full range of evidence available. In, for example, a study of the history of the sheep, bones, skin, contemporary illustrations and documentary evidence must all be considered. Each of these has a particular contribution to make. In this context Ryder (1966, 1969) has studied the evolution of the fleece characteristics and breeds of sheep from the characteristics of the hair follicles preserved in parchments and a broad survey of the history of livestock husbandry in Britain is provided by Trow Smith (1957, 1959). One point to emerge from this study is the extent of the droving industry and the affect this would have on any attempt to study regional livestock characteristics from contemporary deposits.

In many societies, past and present, religious practices, prohibitions and social prejudices exerted a profound influence on husbandry and diet. A full appreciation of the many possibilities and variations of livestock use can only come from a wide reading of historical and ethnographic sources. A valuable introduction to the ethnographic evidence is provided by Cranstone (1969).

Site Examples

The way in which the data from bones are integrated to establish patterns of economic activity is best illustrated by specific examples. Two archaeological sites have therefore been selected for detailed study. The examples illustrate sites of different types and demonstrate both the strong and weak points of the study procedures for determining the economy of a site. The interpretations of the evidence argued in each case are personal ones and do not exclude other views of the evidence. For practical reasons these examples are taken from my own work. The reader should, however, examine other published studies which deal with different sites and other aspects of bone studies. A useful study of the food economy of a prehistoric group illustrating the information that can be obtained from a prehistoric site is that of Parmalee (1965). The variations in butchery practice between sites have not been systematically surveyed but the studies of White (1952–55) and Guilday, Parmalee and Tanner (1962), in North America, are of particular interest and provide a useful link with studies of the tools used. Bone was an important raw material for tools, weapons and playthings and Parmalee (1959) and Guilday (1963) describe the use to which some bones have been put. The studies of Brothwell and Brothwell (1969) and Clark (1965) indicate the type of survey that can be made once sufficient basic site studies have been completed. These, surveying ancient food, and the economic basis of prehistoric Europe are useful reminders of the importance of a systematic synthesis of available evidence as a means of defining outstanding problems and of reviewing achievements, techniques and inadequacies.

Site 1. A Saxon Farm, Whitehall, London

In the course of extensive modifications to existing buildings in the Whitehall area of London around the Treasury Building and Downing Street, Mr Michael Green was able to excavate a number of timber structures, ditches and rubbish pits, the character of which indicates a settlement site of the late ninth century A.D. The settlement has been identified on structural features as a farmstead and this view is supported by the bones found. A considerable quantity of bone was recovered from these dated deposits in the manner laid down in the early chapters of this book and this was subsequently studied by the writer.

The primary analysis of the bones was made in groups as excavated, namely the contents of five pits (one of which was an abandoned grain silo), the filling of a section of ditch and a layer of refuse overlying the floor of the main timber building.

The first point to emerge from the analysis is the small number of species involved and their relative numbers. This is set out in Table 4. The bones of deer were found in only two deposits, in one there was only a single limb bone and in the other a fragment of antler. Only a few fragments of horse bones were found, all of which were in the refuse over the floor. A few dog bones were found but only in two deposits. Bird bones, too, were very few, there being fifteen in one deposit and five in another.

Table 4. Treasury Site: the minimum number of animals present

Deposit ref.	Horse	Cattle	Sheep	Pig	Dog	Deer	Bird
A	0	11	10	7	0	0	P
B	2	20	26	12	1	1	0
C	0	12	9	2	1	1	P
D	0	2	3	1	0	1	0
E	0	3	3	3	0	0	0
F	0	6	0	0	0	0	0
G	0	3	3	1	0	0	0
TOTAL	2	57	54	26	2	3	P

P = present.

9. THE INTERPRETATION OF BONE EVIDENCE

The minimum number of animals of each species was determined for the principal bones of the body in each deposit. The results, which are given in Tables 5–7 are represented in graphic form by Figs 19–21 for the whole site.

Table 5. Treasury Site: the minimum number of sheep determined from the principal bones in each deposit and from the whole site

Bone	Deposit							
	A	B	C	D	E	F	G	Total
Cranium	3	3	1	2	1	0	1	11
Mandible	9	24	4	3	3	0	0	43
Scapula	5	19	2	0	2	0	0	28
Humerus	4	15	6	1	1	0	2	29
Radius	9	21	8	0	2	0	3	43
Ulna	3	6	2	0	0	0	1	12
Metacarpal	5	26	6	1	0	0	0	38
Pelvis	10	0	0	1	2	0	1	14
Femur	2	4	1	0	0	0	0	7
Tibia	7	17	8	1	2	0	0	35
Calcaneum	0	0	2	1	1	0	1	5
Astragalus	0	0	0	0	0	0	0	0
Metatarsal	10	20	9	1	1	0	2	43

Table 6. Treasury Site: the minimum number of cattle determined from the principal bones in each deposit and from the whole site

Bone	Deposit							
	A	B	C	D	E	F	G	Total
Cranium	6	7	6	1	2	1	2	25
Mandible	10	13	12	1	3	6	2	47
Scapula	11	14	1	1	3	1	2	33
Humerus	4	13	3	1	3	0	3	27
Radius	6	13	5	0	0	0	2	26
Ulna	1	8	3	0	2	0	0	14
Metacarpal	8	20	7	0	2	5	1	43
Pelvis	5	0	1	1	2	0	1	10
Femur	4	10	6	0	0	0	3	23
Tibia	5	9	2	0	1	0	0	17
Calcaneum	10	10	5	1	0	0	0	26
Astragalus	6	10	3	2	1	0	0	22
Metatarsal	11	16	5	0	1	0	3	36

126   THE STUDY OF BONES FROM ARCHAEOLOGICAL SITES

Table 7. Treasury Site: the minimum number of pigs determined from the principal bones in each deposit and from the whole site

Bone	Deposit							
	A	B	C	D	E	F	G	Total
Cranium	4	6	2	0	0	0	0	12
Mandible	3	6	1	1	0	0	0	11
Scapula	1	6	0	0	3	0	1	11
Humerus	1	6	0	0	0	0	0	7
Radius	0	11	2	0	0	0	0	13
Ulna	2	9	2	0	1	0	0	14
Pelvis	2	0	0	0	1	0	0	3
Femur	6	12	1	0	2	0	1	22
Tibia	7	8	0	1	1	0	0	17
Calcaneum	1	1	0	0	0	0	0	2
Astragalus	3	0	0	0	0	0	0	3

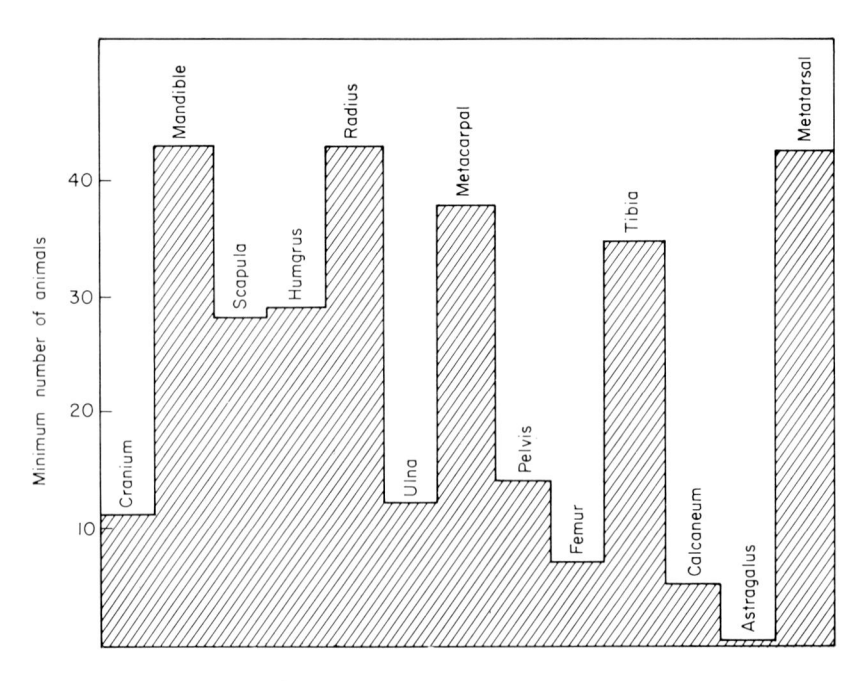

Figure 19   Treasury site. The minimum number of animals determined from each of the principal bones of the sheep.

9. THE INTERPRETATION OF BONE EVIDENCE   127

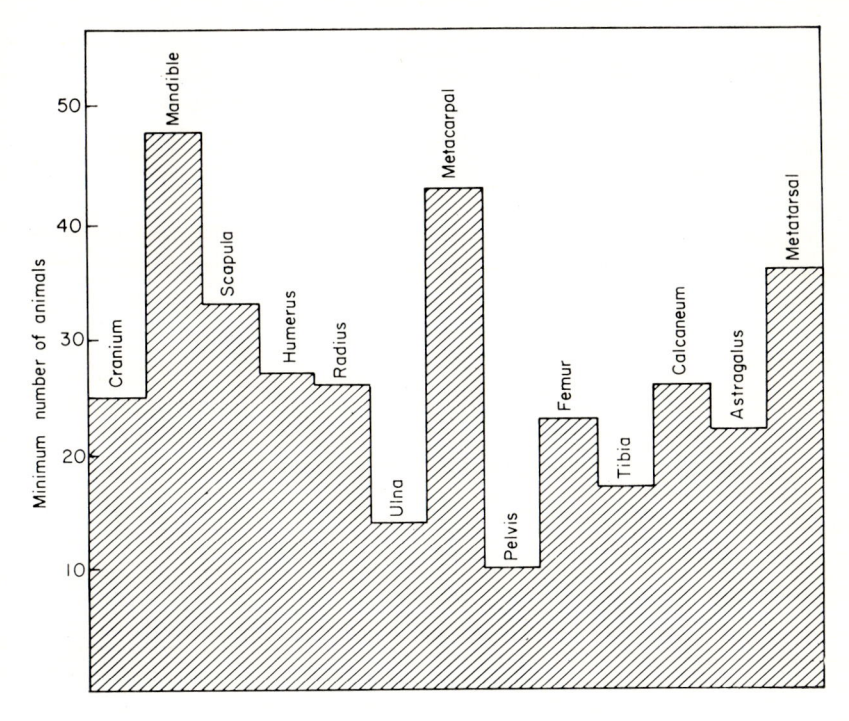

Figure 20 Treasury site. The minimum number of animals determined from each of the principal bones of the cattle.

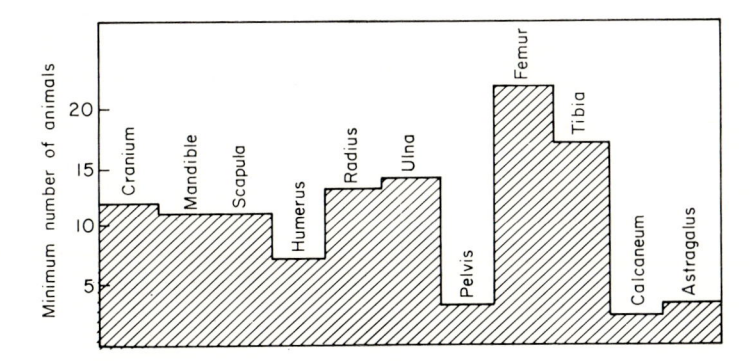

Figure 21 Treasury site. The minimum number of animals determined from each of the principal bones of the pigs.

The fused or unfused condition of the epiphyses of limb bones were recorded in order to determine the age at death of animals represented by these bones. The information was recorded and analysed as the percentage of each bone fused or unfused. (Now this age determination procedure would be applied to the fragments comprising an individual animal so that for each individual identified age and sex could be specifically assigned.)

On the assumption that the sequence of fusion of the epiphyses of a species or a population has remained constant over the centuries these figures can be displayed on a time scale determined from modern stock (or any earlier data) to illustrate the general structure of the husbandry. When using the percentage method as many specimens as possible must be included to obtain significant results. In attempting this problems arise since the age range over which fusion occurs in one bone may overlap the range of another. Further some bones representing a given range may be too few to provide a statistically significant percentage. To illustrate this Table 8 has been compiled to show all the age ranges encountered amongst the sheep on the Treasury site. The figures taken from Silver (1969) are for modern stock. It is clear that only three of the

Table 8. Treasury Site: age criteria for sheep

Age (months)	Bone and epiphysis		No. fused	No. unfused
10	Humerus d.		43	4
	Radius p.		24	4
		TOTAL	67	8
18–24	Metacarpal d.		19	21
	Tibia d.		26	7
		TOTAL	45	28
20–28	Metatarsal d.		13	5
30–36	Calcaneum		6	2
36	Radius d.		15	11
	Ulna		6	2
		TOTAL	21	13
36–42	Femur d.		2	0
	Tibia p.		0	3
		TOTAL	2	3

d = distal; p = proximal.

six groups are of reasonable size, and that some of the six groups overlap each other. In the final analysis it was therefore necessary to take into account only the largest groups and to eliminate those which overlapped. The final analysis of the age data of the sheep must therefore be made in three groups, namely animals aged more or less than 10 months, 18–24 months and 36 months according to whether the appropriate epiphyses have fused or not. These groups, for the sake of brevity, are defined by the age at which the definitive features occur in modern stock. The age structure is then a simple percentage calculation as set out in Table 9. A similar analysis was

Table 9. Treasury Site: age structure of sheep

Age at fusion, 10 months

No. fused, 67; No. unfused, 8; Total 75.

% fused = $\frac{100}{75} \times 67$ = **89·3%**

% unfused = 100—89·3 = **10·7%**

Age at fusion, 18–24 months

No. fused, 45; No. unfused, 28; Total, 73.

% fused = $\frac{100}{73} \times 45$ = **61·7%**

% unfused = 100—61·7 = **38·3%**

Age at fusion, 36 months

No. fused, 21; No. unfused, 13; Total, 34.

% fused = $\frac{100}{34} \times 21$ = **61·8%**

% unfused = 100—61·8 = **38·2%**

made of the bones of cattle and pigs and the findings are presented in summary form in Tables 10 and 11.

Such information illustrating the age at death of the animals is particularly amenable to graphical display. The most convenient format is that of the histogram, preferably a double one showing both the percentage killed in a given age range and the percentage killed before a given age. This is depicted for the three domestic species in Figs 22, 23 and 24.

Ideally one wishes to determine the sex of the animals killed in a given time range but this can only be done where age and sex

Table 10. Treasury Site: age criteria for cattle

Age (months)	Bone and epiphysis	No. fused	No. unfused
12–18	Humerus d. Radius p.	33 (100%)	0
24–30	Metacarpal d. Tibia d.	36 (75%)	12 (25%)
27–36	Metatarsal d.	20 (83%)	4 (17%)
36–42	Femur p. Calcaneum	26 (72%)	10 (28%)
42–48	Humerus p. Radius d. Femur d. Tibia p.	38 (81%)	9 (19%)

d. = distal; p. = proximal.

Table 11. Treasury Site: age criteria for pigs

Age (months)	Bone and epiphysis	No. fused	No. unfused
12	Humerus d. Radius p.	9 (56%)	7 (44%)
24	Metacarpal d. Tibia d.	14 (54%)	12 (46%)
42	Humerus p. Radius d. Ulna Femur p. Femur d. Tibia p.	12 (23%)	41 (77%)

d. = distal; p. = proximal.

data are related to the fragments which comprise each individual animal. This was not done for this site. Bones were measured and the dimensions analysed in order to reconstruct the sex structure of the sample as a whole. Regrettably the number of measurable bones was extremely small for a sample of this size and the results were not conclusive for the dimensions measured. The dimensions studied were of unknown significance for the detection of sexual trimorphism and were principally measures of the width of the

9. THE INTERPRETATION OF BONE EVIDENCE

ends of long bones. Histograms of the frequency of a given measurement for the sheep produced a distribution similar to a normal one, therefore suggesting a homogeneous sample. There were far fewer measurable bones of cattle than sheep. The histograms brought out no certain sex structure except that most showed a single specimen

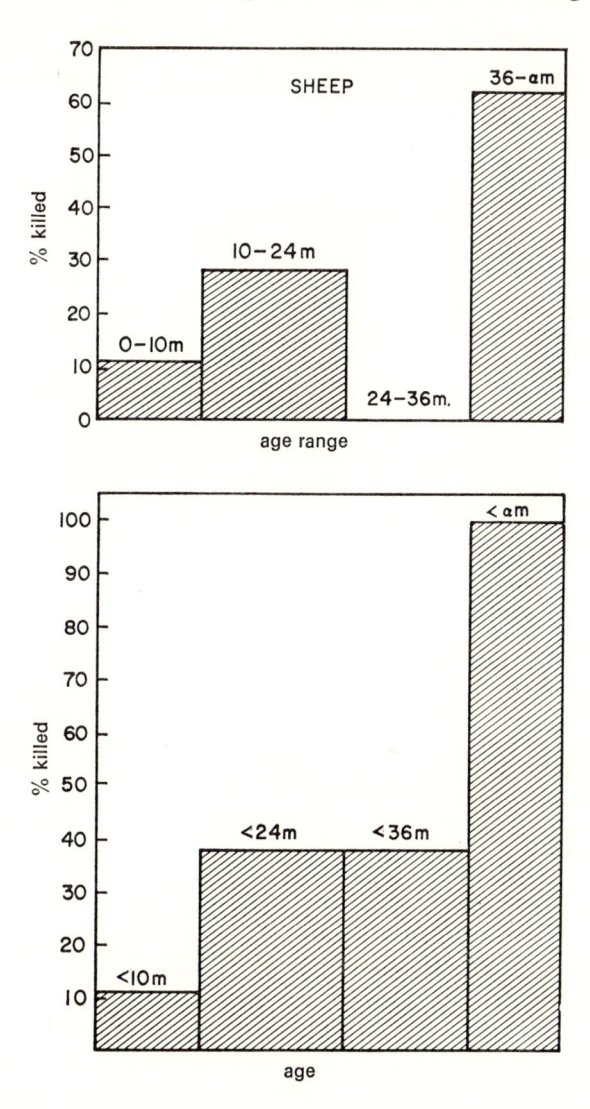

Figure 22 Treasury site. Age at death of sheep as judged by epiphyseal fusion. (Top; percentage killed in age range. Lower; percentage killed before the given age.)

to be much larger than the others. Calculation showed that this specimen was between two and three standard deviations from the mean. This probably represented a sexually intact male, i.e. a bull. The bones of pig were still fewer and the results quite inconclusive. In interpreting the evidence set out above we must first turn to

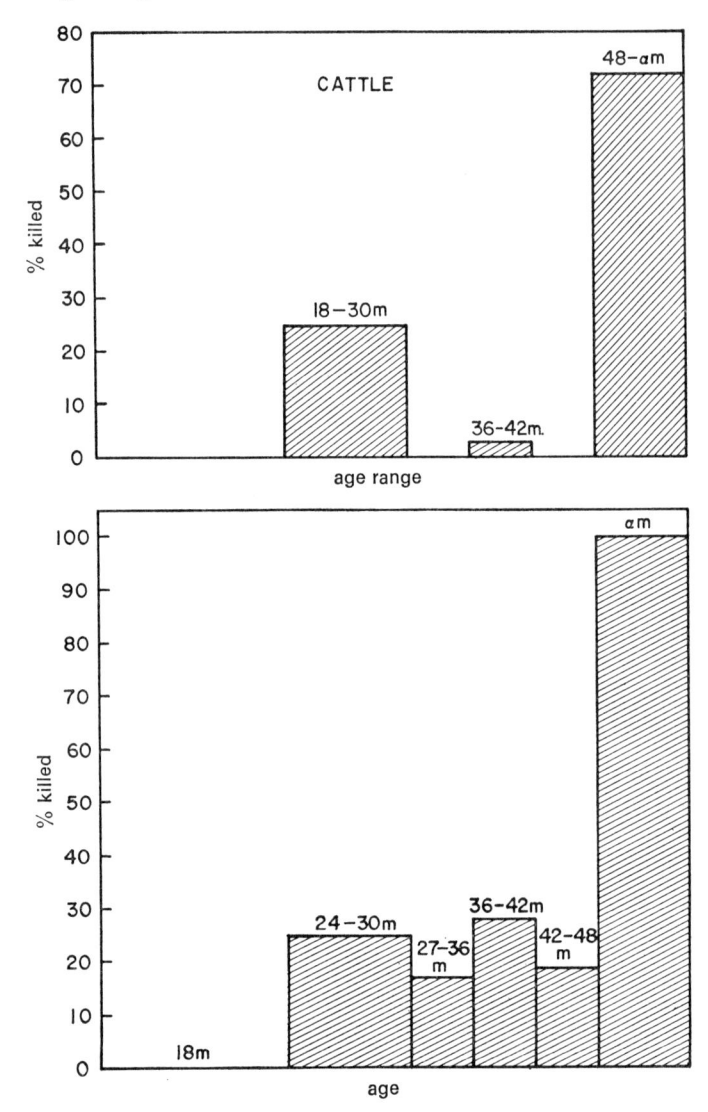

Figure 23 Treasury site. Age at death of cattle as judged by epiphyseal fusion. (Top; percentage killed in age range. Lower; percentage killed before the given age.)

9. THE INTERPRETATION OF BONE EVIDENCE

the summary provided by Table 4 to consider the basis of the animal food supply. Examination of the bone debris—its fragmentation, butchery marks, etc—showed that it should be regarded largely as food refuse. We may therefore now consider the role of each species in the food strategy. The only numerically important species are the domestic cattle, sheep and pigs in that order. As we are con-

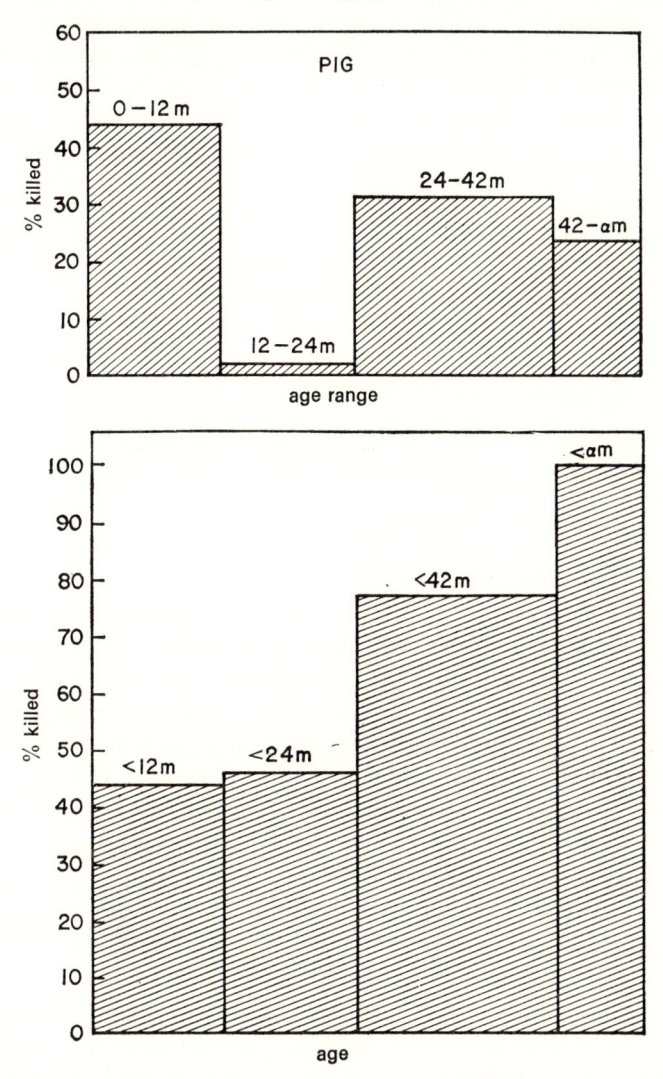

Figure 24 Treasury site. Age at death of pigs as judged by epiphyseal fusion. (Top; percentage killed in age range. Lower; percentage killed before the given age.)

cerned with the diet it is not enough to consider just the numbers of each species. The actual food contribution is what must be considered and this will include not only meat but offals, paunch, etc. In order to do this we must make some assessment of the relative weights of the different species in terms of so many pounds of meat and offals or live or dressed carcass weights. If that is not possible, or it is considered inappropriate to use an actual weight, then a factor may be used; x sheep $= y$ cattle $= z$ pigs, etc. Although there is in many cases a strong justification for using total body weights it is often easier to obtain standardised comparative data on dressed carcasses. In Table 12 I have estimated the probable dressed carcass

Table 12. Treasury Site: carcass yields

Species	Minimum no. animals	Sheep equivalent (S.E.) value of species (units)	S.E. units present	Estimated dressed carcass Wt (lb)	Calculated carcass yield (lb)
Cattle	57	12	684	300	17,100
Sheep	54	1	54	25	1,350
Pig	26	2	52	50	1,300
				TOTAL	19,750

weight for the type of stock on this site. How correct the figures are is unknown but proportionately they are probably about right. From these figures we can calculate the relative quantity of meat from each species and the approximate total weights involved.

The calculations give a figure of 19,750 lb of meat from the three domestic species, of which the cattle account for 17,100 lb. Beef is thus the most frequently eaten meat and cattle are the staple meat producers. From the other evidence we must now establish whether there is any selection of joints or indications of specialised meat production for home consumption. Examining the age range killing graph for cattle (Fig. 23) it will be seen that on our modern time scale some 25% are killed quite young in the 18–30 month range whilst the remainder bar two are killed at 48 months or more. The implication here is of the selective killing of some younger but grown stock whilst the remainder are left to maturer years. It is possible that this younger group represents castrated males. The absence of calves killed is noticeable. For sheep the selective killing

9. THE INTERPRETATION OF BONE EVIDENCE

of grown young stock and older animals is indicated by the data in Fig. 22. The most dramatic evidence for selective slaughter is provided by the pigs, with peaks in the 0–12 and 24–42 month groups (Fig. 24). Thus selective killing on an age-related basis was practised on these animals. It is likely, but cannot be demonstrated on the evidence available, that sex groups were also involved with age. It is not, however, implied by these graphs that there was any seasonal pattern to this slaughter; nevertheless it should be borne in mind that if meat were to be preserved in some way such as salting, then the selected animals would be killed at the most favourable season.

The next step is a very difficult one. Can we extrapolate from the food debris to the farming economy and is there any evidence of trade in live- or dead-stock? All too often food debris is used without justification to present a picture of the husbandry. The diet can be a cross-section of the husbandry but it can equally be a cultural artefact. On a farm site of this date close to a major settlement (the City of London) it is very likely that the husbandry was commercial and that the food refuse would not present a true picture.

We know that animals were eaten here and we must also assume that being a farm some animals were killed and sold as carcasses whilst others were marketed on the hoof from here. Although it is well known both from archaeological evidence and documents that animals were marketed on the hoof, often over long distances, we cannot demonstrate this at the site of production. We can, however, determine whether dressed carcasses or parts thereof were dispersed from the site. Certain parts of the body have little food value and are not marketable. These waste items and associated bones are normally removed at the point of slaughter when the carcass is dressed. Provided they are not sold for further processing these bones should be found with the slaughter's refuse. Thus the possibility of the sale of dressed carcasses or particular joints can be studied by reference to the figures showing the number of animals determined from the principal bones of the body. In the case of the sheep (Fig. 19) the waste is represented by the metacarpal and metatarsal bones which, with the tibia and mandible, are the most frequently occurring items. The total from the scapula and humerus fall somewhat short of the maximum whilst the femur comprises less than a quarter of this.

Here there is a good case to be made for the dispersal of selected joints—leg of mutton and shoulder rather than the dispersal of whole dressed carcasses. However, another explanation could be that the dogs consumed some of these items. Experiments with our own dog on such joints leaves little doubt that they could be totally destroyed. Against this must be set the fact that very few fragments show evidence of having been eaten and the larger cattle marrow bones which are not totally destroyed by dogs show no evidence of having been gnawed. If the distribution is artificial and not a reflection of marketing the limits of fragmentary identification could produce some bias against the femur, though not so easily against the other items. I do not, however, believe that the methods employed could lead to an error of the magnitude implied by the figures.

In the case of the cattle, the mandible, metacarpal and metatarsal are the most frequently occurring bones and the figures from the other forelimb bones are fairly similar to each other, as are those of the other hind limb bones. We are again faced with the dilemma of dispersal as opposed to selective underestimation. Where the ends of the bones are preserved as on this site, it is not possible to seriously underestimate any one bone. The fact that other bones such as the astragalus and calcaneum, which are not greatly fragmented in carcass preparation, complement the numbers determined from marrow bones suggests very strongly that we are dealing with a genuine economic activity and not an analytical artefact.

The pigs present a more homogeneous picture implying no discernible disposal except perhaps of the foot bones. It is possible that we are dealing here with the joints of meat that may have been salted, a process from which the trotters may have been excluded, perhaps being cooked up at the time; or maybe the dogs did have them.

A good case can be made for identifying this bone refuse as a mixture of domestic food and commercial debris, further, for a given animal killed, some joints were eaten here, whilst others were dispersed. Therefore we have good reasons for considering that the food debris is related to commercial activity and it represents a cross-section of the husbandry. Further support for this view comes from the graphs showing the age at killing. The picture presented is one of rational husbandry, a multi-purpose economy which one would consider reasonable for this site.

The cattle are killed in two groups—young, but grown animals

9. THE INTERPRETATION OF BONE EVIDENCE

and an older age group. This implies the selective removal of younger stock at an economic carcass weight and these animals must be regarded as having been reared primarily for food purposes. It is likely (though unproven) that these animals were castrated males, i.e. steers. The older group can be taken as representing stock for cropping; that is, they formed the breeding herd, may have provided milk and were probably used for draught work. There is a strong implication from their numbers that cattle were the most important livestock kept and this may in part be related to the habitat of the low lying Thamesside area of Westminster.

The three groupings of the sheep are also significant. The first group is difficult to explain on purely economic grounds. It could indicate a certain choice of young sheep but these would have provided very little meat in their first year. The figure is, however, very close to the normal losses encountered in modern sheep production in the first year of life and therefore it is possible that these were animals that died and were eaten or that they were sickly, weak or late-born lambs that were killed because of this. The middle group is explicable as the extraction of prime mutton probably from castrates. The older group, as with the cattle, implies the cropping component with breeding stock, wool production and probably also milk. The quality of wool produced by ewes, lambs, castrates and rams in a given breed varies both in quality and quantity between them, but there is no objection to the keeping of castrates for wool if the circumstances permit. Castration is practised primarily to modify growth in favour of certain carcass types (at least in modern husbandry) but in ancient times perhaps more importantly in order to obviate the problems associated with the sex drive of the intact male. If resources are slender the ewe with its lamb and milk is the most productive animal to raise and care for. The numbers of the sheep are quite small; the bones imply that sheep were far less important than cattle and therefore that the sheep flock at any one time would have been small. In assessing this one is, however, up against the age-old problem of the duration of occupation indicated by the debris and how much of it was dispersed elsewhere. Although the pigs are relatively few in number the picture is extremely interesting. Over 40% of the pigs were killed young with a big gap between these and the older age group. The breeding biology and fecundity of swine in general and their ability to develop on a very varied diet makes it possible for a high proportion of young animals to be slaughtered without prejudice to the breeding stock. Whether they were eaten fresh, salted down or smoked we do not know. In late winter the ruminants would be in their poorest condition owing to the lack of any quantity of natural forage. The ability of the pigs to find and eat a variety of foods would result in their being in better condition than other stock at this time of year. The value of fresh meat at a time when the farmer would be loth to slaughter ruminants in such poor condition would be of great psychological value. In peasant agriculture the pig probably provides flexibility in the husbandry pattern, and a source of fresh meat at times unfavourable for the slaughter of other stock. The enlargement of breeding herds or changes in strategy in the husbandry of cattle and sheep take several years to achieve. In the event of disease or other catastrophe it might be impossible to adjust the herd structures rapidly enough to maintain or increase the herd. The pig with its large and frequent litters and early puberty provides a very valuable safety margin in primitive husbandry, especially where this is specialised towards cattle rearing.

From the evidence available the inferences that can be drawn are fairly clear. The husbandry is product-orientated, that is, it proceeds on rational grounds and is conducted at a competent level. Cattle rearing seems to dominate. Presumably there was arable cultivation at least to provide winter fodder. In connection with arable cultivation we must remember the value of livestock in manuring the land and their value as draught animals. Whether the stock were fed or stalled part of the year we do not know. The farm with its palisade would, however, have been ideal to contain the stock at night, to prevent them straying but, perhaps more important on ninth-century Thameside, to prevent their being stolen. The success of the husbandry would, however, suggest some provision of winter feed.

Although this simple rural picture is not what one has popularly been led to believe about our Saxon ancestors, just as the literary Camelot does not resemble the archaeological reality, it is in fact the same picture that we get from the later Domesday records.

Site 2. Medieval Workshop Debris

In 1960–61 excavations were carried out by Mrs Charmian Woodfield in Well Street, Coventry. The report on this excavation along

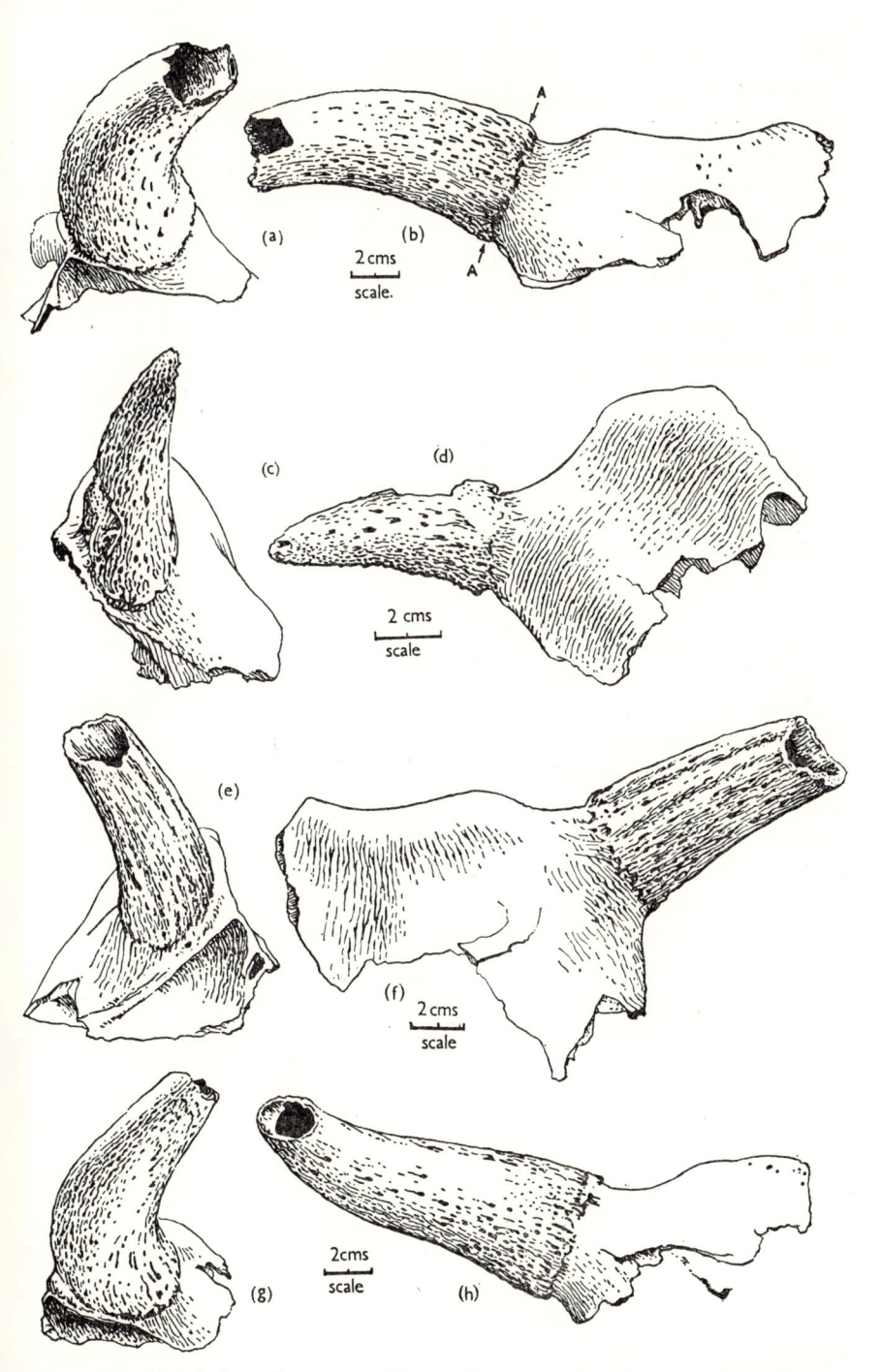

Figure 25 Morphological varieties of bovine horn cores found in a medieval deposit at Coventry.

with that on the bones has been published (Gooder, Woodfield and Chaplin, 1966).

Comparatively little bone was recovered from the site and interest centres on two collections of material (referred to as groups 1 and 2) which are associated and contemporary and can therefore be amalgamated at this stage. This group comprised thirty-seven horn cores and about half a dozen other fragments of bone.

The group was of interest both in terms of the origin of such a large number of horn cores (the deposit was more extensive than excavated) and the information that they could give about the different types of cattle that found their way into fourteenth-century Coventry.

On the basis of the form of the horn core and the shape of the cranium the horn fragments were assembled into different groups. Each group, comprising a basic homogeneity of form in both the shape of the skull and the form of the horn (Fig. 25), was measured and the results were plotted as a frequency diagram (Fig. 26) for each morphological type. The reason for this was that it was believed that the morphological groups represented distinct breeds or types of stock rather the different sex groups of a single breed. It was therefore necessary to assess the sex structure of each group. Because of their fragmentary condition the only dimension that was generally applicable was the circumference of the base of the horn core. In modern horned Bovidae and almost all other horned ungulates the horns are sexually dimorphic, large horns being associated with the males. Only two morphological groups were present in any number and when plotted the measurements clus-

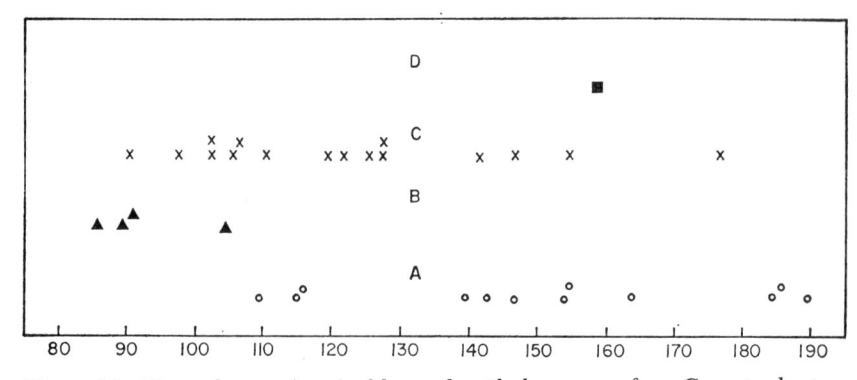

Figure 26   Circumference (mm.) of base of cattle horn cores from Coventry by type.

9. THE INTERPRETATION OF BONE EVIDENCE

tered into groups. In type A stock the distribution is trimorphic and in type C (Fig. 26) tetramorphic. This latter distribution is not to be expected in a herd sample and it is very likely that this is just an extreme example, the sample being too small to cover fully the range of variation encountered. Equally it could be a different type of animal whose horn conformation happened to coincide with that of type C animals. Types B and D are not well represented in the collection but stand as very distinct forms. The distribution of the measurements suggesting the presence of males, females and castrates, in these groups A and C lends support to the view that the groups represent different types of animal.

The resemblance of these horn cores to those of modern breeds of cattle is at once apparent and it would be very easy to refer to these groups by the breed names which they resemble. Such temptations must, however, be resisted as the origin of many of the modern breeds is quite late and often little known. Further, breeds are characterised by a great many features such as milk yields and growth characteristics whilst coat colour and cranial morphology are irrelevant characteristics to the animal breeder interested in the economics of livestock farming. It may happen, but cannot be relied upon, that such characteristics as horn form and coat colour do remain in the selection process. To imply that there is a connection over many centuries with a modern breed on such a variable character is to defeat the whole object of the exercise which is to establish a factual basis for the study of the history of cattle types.

In the report, therefore, I feel that one can do no more than describe the characteristics of each group of horn cores and to give a name to the groups. In this case they are designated Coventry types A–D. It will thus be possible to discuss these and other examples without the use of loaded or suggestive terms. Once an adequate series of located and dated types has been established it may be possible to reconstruct with some accuracy the way in which regional livestock types developed and how they were used to develop the modern breeds.

A deposit of any one type of bone requires explanation. We were therefore asked to establish what these horn cores represented. The deposit was in fact much more extensive than could be excavated and obviously involved many hundreds of horn core fragments and very little else. Although the cores themselves are damaged this is probably due to their friable condition and it is unlikely that they were broken when discarded. To explain how these fragments could have accumulated it is necessary to consider the manner in which a carcass might be disposed of to produce the observed result.

The carcass of an ox may be disposed of from the slaughteryard as follows. The dressed carcass excluding the skull and often the feet would go to a butcher, the hide to a tanner, the gut, offal and the intestines for sausage skins, etc, to a butcher. The waste bones —the skull after removal of the tongue and maybe brain, could, with the feet, be sent for boiling into glue or fat or they might be further divided up, the cannon bones, for example, being used for the manufacture of pins and other objects, and the horns sent to a horner, the remainder of the skull then going for glue. The horner does not use the bony core but only the outer keratinous sheath (the true horn) which is stripped off, the core then being of no further use is classifiable as debris from a manufacturing process. In a slaughteryard skulls and feet would only be found if they were not being sold, and bones from any of the domestic animals would almost certainly be present, not just those of one species. If the skulls were not being disposed of they would not have been carefully cut up. The present horn cores have been systematically detached from the rest of the cranium and there is no trace of the other skull fragments. It can therefore be concluded that the horns themselves were of primary interest and that they had been collected together with a process in view. The only process which would require such a raw material is the processing of horn which we know to have been an important commodity in medieval times. The keratinous horn may be separated from the bony core by maceration, after which the outer horn sheath can be pulled off or it falls away. This is then processed into a variety of articles.

In this instance the deposit appears to consist almost exclusively of this debris. Another site that I have examined yielded a similar collection of horn cores but this time with an admixture of domestic debris. It is therefore important to remember that it would be very easy for other debris to get mixed in with processing refuse and one must always be on the alert for inconsistencies which might indicate a multiple origin of the debris. One should also bear in mind when arguing on the above lines that several activities may have been carried on in close proximity to each other, if not by the same concern.

10
Animal Remains as Indicators of Past Environments

Terrestrial Mammals

Animal bones were probably the first biological material to be used by archaeologists to assess the environment of early man. The identification of animals such as elephants and hyenas on northern archaeological sites was of considerable significance in establishing that the climate had changed since man first appeared. The association of bones and artefacts in cave and river deposits left no doubt that these animals were contemporary with Stone Age man.

Traditional Pleistocene chronology is based on the alternation of warm and cold phases, each with its indicator species. Thus the mammoth and reindeer indicate cold conditions and red deer, bison and elephant warmer conditions. In general, this simple division with various refinements has served quite well for northern Europe. The application of newer environmental techniques has, however, shown that these broad faunal categories are too imprecise for the detailed study of the Pleistocene. In part this imprecision is due to a lack of ecological data about contemporary species and faunas. It is therefore necessary to discuss the way in which animal communities and species can be used as indicators of the environment and also as a means of relative dating. The same principles and problems that will be discussed in connection with Europe are also applicable to other regions.

The terrestrial fauna is best considered in two groups. One group covers the larger mammals, the other the smaller. This division is arbitrary and the critera can be adjusted according to one's needs. In the present context I distinguish between the large mammals, which provided prehistoric man with the bulk of his animal protein, and the small mammals, such as mice and voles, which are more influenced by microclimates and have smaller home ranges than the larger ungulates and were probably less often used as food.

In comparison with the mammalian fauna of mainland Europe that of Britain is relatively impoverished. Of 101 indigenous terrestrial mammals in Europe only 39 are also found in Britain. The composition and origin of the British mammalian fauna is summarised in Fig. 27. A full species list is given by Corbet (1964).

indigenous survivors	introduced	extinct
39	19	7

Figure 27 Structure of the British post Pleistocene terrestrial mammalian fauna.

Some species such as the brown bear, wild pig, elk and reindeer became extinct in Britain during post-Glacial times. The distribution of many mammalian species in Europe suggests that formerly they were more widespread and that populations now isolated may once have been continuous. Faunal lists from post-Pleistocene sites in Europe confirm this. Historical records also indicate that many large species are now more numerous than they were a few centuries ago even though they are not so widely distributed or abundant as they were in the pre-agricultural phase of the post-Glacial. This is the result of conservation activity by sporting interests and extensive re-afforestation in the last century. In the early eighteenth century roe deer were almost extinct in Britain and survival nuclei were established with stock from Europe. Since the extension of state and private forestry from the 1920's roe and other deer have spread over much of the country. Similar colonisation movements have occurred with roe and elk in Scandinavia.

The fate of the once numerous and extensive herds of game animals familiar to Mesolithic man is very pertinent to our attempts to interpret former environments from fossil faunas. It is known from vegetation studies that the post-Glacial is a period of rapid ecological change that might alter the composition and distribution of the faunas. Equally drastic changes had, however, occurred many times during the Pleistocene without impairing the fauna. Climatic change accounts for only a few of the major species shifts during the post-Glacial. Reindeer migrated northwards and horses east-

10. ANIMAL REMAINS AS INDICATORS OF PAST ENVIRONMENTS

wards with the habitats. Throughout the world the post-Glacial is marked as a singular epoch by the explosive economic and cultural evolution of man and the resultant modification of the habitat. It would appear that the present distribution and status of game animals in Europe is a consequence of this. Despite the resurgence of some species many areas will never be recolonised. Species such as the musk ox and bison will remain scarce whilst others like the aurochs (*Bos primigenius*) have disappeared for ever.

Similar drastic environmental and faunal changes have occurred in recent centuries in North and South America, Australia, China, India and elsewhere by similar means. It is only the rate and scale of the change that is different.

From the faunal record of middle Eastern and Mediterranean sites it is clear that similar but more drastic changes occurred here quite early. The motives for conservation were not then developed and the losses have been almost total. The Sassanian kings of the second millennium B.C. who hunted the Mesopotamian fallow deer would probably have killed with their entourage in a day a "bag" equal to the total present world population of this deer. Other favourite prey of these kings were lion (now extinct) and wild pig (scarce).

Thus throughout the post-Glacial period the ecology of the game species has been increasingly dictated by man. For this reason I do not believe that we can apply the ecological requirements of present populations to those of the past with sufficient precision to use game mammals as a primary source of environmental data. This restriction is a matter of precision rather than total inadequacy. Modern populations still have their ecological needs, optimum habitats and limitations which can be investigated. In some cases these will probably be comparable with those of some ancient populations. An examination of the fauna associated with the major European habitats will illustrate the sort of deductions that can be derived from a study of present conditions.

In northern and eastern Europe there are five major habitats which appear to have been important in the evolution and localisation of species during Pleistocene and post-Glacial times. These are: the Tundra, Boreal (northern coniferous) Forest, Deciduous Forest, Steppe and Alpine zones. Their present distribution is shown in Fig. 28. However, these zones are not clear cut; large areas are transitional in character and the zonation occurs not only horizontally but vertically and is also modified by local soil conditions and microclimates.

In these habitats are associations of species (faunas) which have evolved (or adapted) together and exist in dynamic relationship to

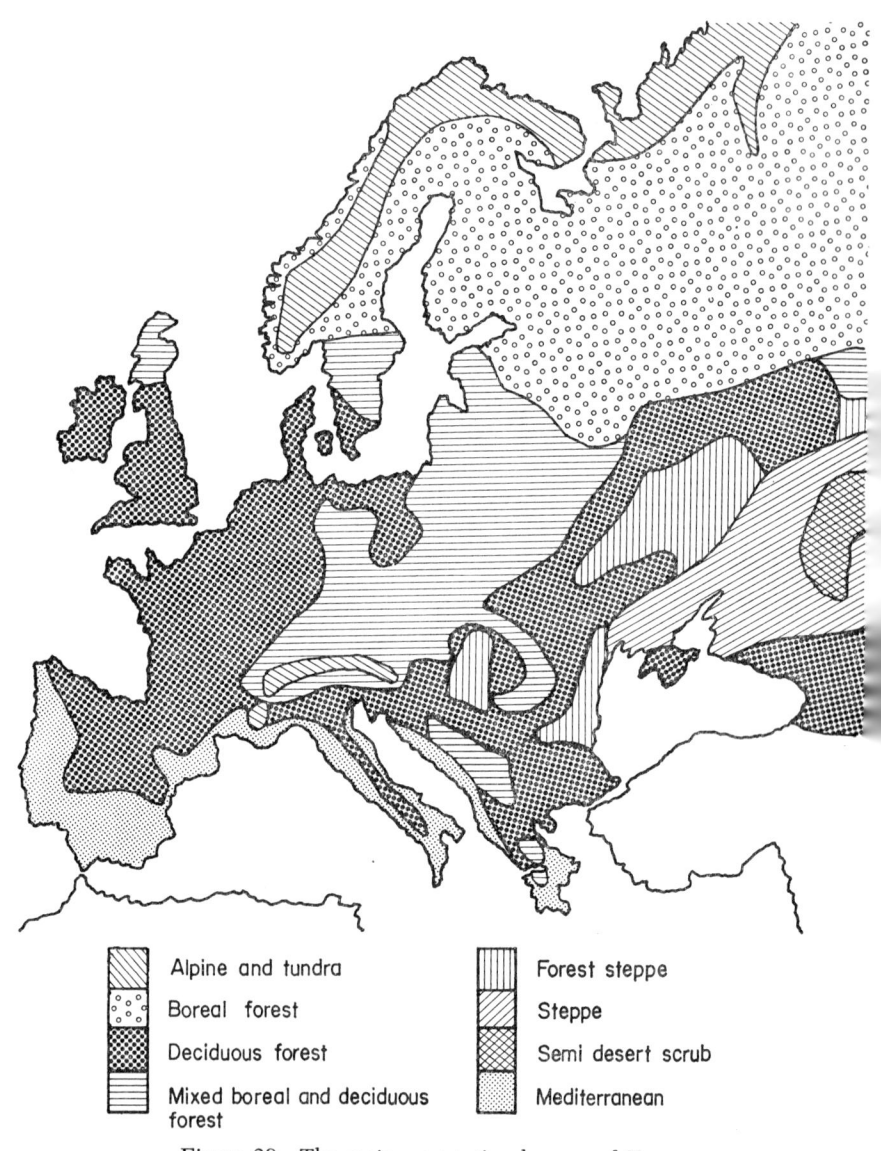

Alpine and tundra
Boreal forest
Deciduous forest
Mixed boreal and deciduous forest
Forest steppe
Steppe
Semi desert scrub
Mediterranean

Figure 28 The major vegetational zones of Europe.

the habitat and to each other. Adaptations to this habitat do not necessarily restrict a species though often an adaptation will establish this as the optimum habitat. These adaptations can be behavioural, physiological or anatomical. Many of them are not to the total ecosystem but to a part of it. Thus a species may be found in a variety of vegetational zones but occupying the same ecological niche. When studying faunas it is very important to establish the adaptations of each species within the ecosystem.

The dominant adaptation of large mammals in the deciduous woodland and transitional zones is to deciduous trees and dense ground vegetation. It is the ground vegetation which provides the greater proportion of food and occupation at certain seasons may depend on the abundance of only a few plant species. Thus in the deciduous woodlands of southern England during the winter, deer are very much dependent on blackberry, ivy and holly to supplement the dead grasses and adjacent cultivated crops.

The large ungulates found in the deciduous forest are predominantly ruminants. Individuals and small groups are markedly territorial. The coloration is cryptic, protection being by camouflage rather than flight, though escape is frequently along well worn paths. The diet of these animals appears to require variety. Thus both grasses and other ground plants are the main food source, but browse (shrubs, leaves, etc.) is also required. Our captive deer do not normally eat hay, but after a period on leafy green plants they develop a craving for more fibrous food and will then eat hay, woody plants or even cardboard. Given an abundance of woody plants they crave for leafy material. The need appears to be for a balance of nutrients plus roughage to aid rumen function. Food habits change during the year as certain plant species reach stages in their life cycle.

Bison, wild pig, red deer and roe deer occur in these woodlands. Fallow deer also flourish there but although present in the Pleistocene were only introduced to northern Europe in Roman times. Their effect on the indigenous populations is unknown. Within this group one notes from present-day distributions and habitats that both red and fallow deer are to be found in parkland but roe deer appear to flourish only in woodland.

The steppe fauna is now impoverished and represented only by the saiga antelope and a very few wild horses. Formerly it included a rhinoceros and mammoth. The particular adaptations of antelope and horses are to an open landscape. The senses, speed and endurance are well developed to detect and escape predation. The teeth and digestive tract are adapted to dealing with a diet predominantly of coarse grasses and other low-growing plants. In non-ruminants such as horses protracted feeding is necessary, unlike ruminants that can rapidly fill their rumen and retire to cover in order to chew and digest it. The location of the steppe zone in central Europe means that these species must withstand long periods of low temperatures and snowfall in the winter months and hot summers. Thus the physiology of the animal is adapted to this and the pelage likewise, there being a thick winter coat and a thinner summer one. It will be appreciated that we know little of the adaptations of these species to steppe conditions and of this fauna only the saiga antelope is left for further study. In those situations where several members of the evolving fauna have become extinct it is possible that there have been major consequent changes in the range and behaviour of the survivors. A successful introduction may also have a similar effect.

The boreal forest is not suited to the needs of the majority of steppe or deciduous woodland species. This is principally owing to the lack of ground vegetation in mature stands and to the harsh climates to which they have not adapted. In Europe few ungulates have made it their home although in other continents such as North America it is within the range of some species which occur in steppe and deciduous habitats. The woodland reindeer is adapted to these conditions and associated races penetrate the tundra. The woodland reindeer are not confined to the coniferous woodlands but range into the birch scrub associations that succeed the boreal forest altitudinally and northwards. This habitat is also shared with the elk. Both the reindeer and elk occurred in Europe in the late-Glacial period and migrated northwards early in the post-Glacial period. They were extensively hunted by Mesolithic man and this has probably restricted their southern limits.

The tundra zone and the forest margin are also occupied by a race of reindeer adapted to this habitat. One of the most important adaptations is the seasonal migration between forest and tundra. Woodland and tundra reindeer are conventionally distinguished by characteristics of their antlers but whether these forms are valid in prehistoric times is doubtful. Associated with elk and reindeer in cold habitats is the musk ox *Ovibos moschatus*. This species is of great antiquity and is fairly common in Pleistocene deposits in Europe. In the late Pleistocene period its distribution was circumpolar but in post-Glacial times it has become virtually extinct the only populations being in Canada and Greenland. Its extinction in Europe is probably due to hunting by man. There would appear to be no area where the tundra fauna as we believe it to have existed in the late Pleistocene period is preserved.

The Alps are characterised by a vertical stratification of vegetational zones. The alpine vegetational zone—above the tree line—appears not to have been extensive enough or of sufficient significance for any species to have adapted to it. At the present time the chamois (*Rupicapra rupicapra*) and the mountain ibex (*Capra ibex*) are found in the Alps above the tree line, and apart from hunting pressures both flourish there. It seems probable that both of these species are secondarily adapted to this habitat and that the chamois in particular was originally a part of the woodland or steppe fauna. However, the history of both species is obscure and more evidence is needed to trace their adaptations.

From this brief survey it is apparent that the major zones have their characteristic faunas the species of which do not make incursions into the typical habitat of adjacent zones. The areas of contact are the transitional zones where local factors may cause the vertical or horizontal interdigitation of habitats or where one zone slowly gives way to the other. This relationship in Europe is shown in Fig. 29. These transitional zones form a significant percentage of the total habitat. On this basis it is possible from study of the faunal remains to give some indication of the zonal location of an archaeological site.

In attempting to locate a site in this way it is necessary to place it within a present-day vegetational zone and to draw up a faunal list for this and adjacent zones. The nature of the accumulation should be ascertained (i.e. natural, animal or human origin) and the relevant factors influencing the sample should be noted. Finally the particular local conditions should be considered and note made of any topographical features likely to modify the broad faunal spectrum. The faunal sample can then be compared with the zonal one to indicate if it belongs to this association or is characteristic of adjacent or transitional zones.

Although ecological interpretation of any deposit is complex the greatest problems are with accumulations of human origin. Some of the complicating factors affecting accumulations on hunting sites have already been discussed. The character and content of the home midden of any hunting group is unlikely to be representative of the species present, their abundance or distribution in the area nor always of the species and numbers obtained. Thus on prehistoric

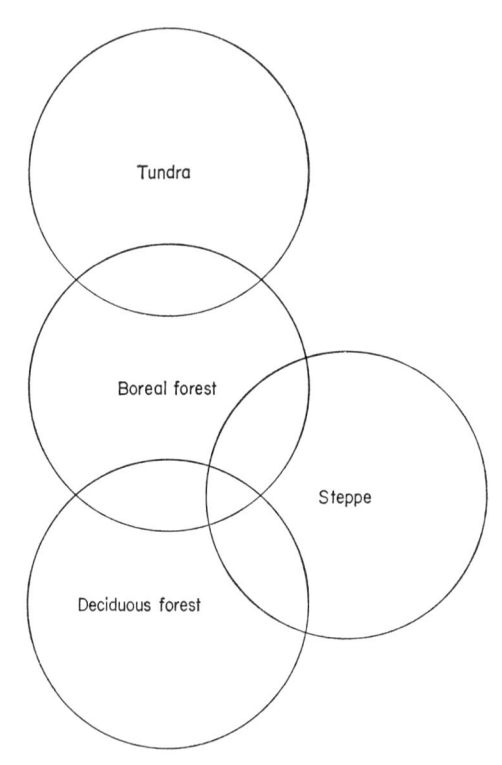

Figure 29 Schematic relationship of the major vegetational zones of Europe during the Pleistocene period.

habitation sites the proportional representation of species is not suitable for determining the ecology of the home range. This can only be assessed according to the presence or absence of species when absence is not due to cultural artefact and occurrence is not anomalous.

The general situation discussed above—that of northern Europe—is applicable to most undulating terrains of similar scale since the home range of individual human groups is unlikely to have included two major zones. Many areas of critical concern to archaeologists do not conform to this pattern of extensive horizontal zonation. These include the numerous mountain and valley zones of Europe and Asia, north and south America, and the coastal mountain ranges of north America and parts of north Africa. In such areas ecological zonation is primarily vertical. Where ranges produce a rain shadow effect, vegetational zones are frequently condensed within very short horizontal distances. Such areas may present a microcosm of the ecology of a continent.

Vertical zonations of vegetation associated with temperature are relatively stable. Those associated with rain shadow effects may be unstable if weather patterns are variable. Short-term effects such as the failure of the rains in a single year are far more destructive to a fauna than long term changes. The coincidence of a late season and high population levels can decimate a herd or cause widespread migrations into adjacent areas.

In mountain terrains the home range of single human groups may span several distinct zones. Thus the zone(s) and species exploited are the subject of a relatively free choice. In areas with a known climatic failure rate there will be years when different faunas are exploited because of short-term fluctuations or migrations. Thus in studying sites in these areas great caution is needed in interpreting fluctuations in the faunal sample.

The habitats of post-Glacial times are of comparatively recent origin. Until Villafranchian times, apart from ancient glacial activities, temperatures were higher and the cold environments of the present day unknown. In the course of the last two million years there has been a progressive fall in mean temperature. Thus at the beginning of the Pleistocene period there were no plant or animal species that had evolved under cold conditions. During the Pleistocene period ancestral species evolved new ecological forms and subsequently species adapted to these conditions. Those that could not adapt became extinct or lived on only in relict areas. Whether considered on a total or individual basis, species evolution in the Pleistocene period was very rapid. The climatic deterioration was progressive and wave-like so that continuous adaptation was necessary. Rapid adaptability was a great advantage, primitiveness rather than specialisation being the key to evolutionary success.

The adapations could be behavioural, physiological and anatomical. Apart from changes in size or dental complexity as in the Proboscidea, skeletal adaptations were not well marked and were largely incidental. From present knowledge of environmental adaptations at the sub-specific level it is clear that many important adaptations will not be reflected in the skeleton. In some cases there may be indirect skeletal effects, although these manifestations may give no indication of the factors concerned. The Scottish and English grey seal populations differ in the frequency of certain minor cranial features but these give no indication of the forces that have led to this or of the other factors that may be associated genetically or incidentally.

The analysis of local traits has scarcely begun to be applied to Pleistocene problems. Ecological interpretations of these traits would be almost impossible to support without other environmental evidence. The techniques used to study the Pleistocene faunas involve the application of arbitrary criteria to distinguish species. These species were often no more than forms in a lineage showing continuous variation and undergoing adaptive radiation. Ecological assumptions derived from studies of a surviving form were then applied to these arbitrarily divided lineages.

The application of recent ecological data to increasingly older populations of the same species ignores their evolution. This approach is valid only if we recognise its limitations and are prepared to test the implicit hypotheses whenever data are available. The application of this technique to faunas rather than species has probably avoided serious errors of interpretation. There is now a tendency to ignore the fauna and concentrate on individual species. At this level modern ecological data (albeit incompletely known) have been applied to marker species. A more profitable approach would seem to be the converse of this. With recent advances in dating and environmental techniques it is possible on some sites to reconstruct the ancient environment with some precision. It is thus possible to concentrate on the study and interpretation of faunas or species within their context. This approach will involve a greater study of local populations of mammals and an emphasis on the investigation of features associated with evolution at the sub-specific level. In this way it will be possible to trace forwards the fate of particular adaptations and in time will indicate the applicability of modern data to ancient populations. Such an approach will be of great value to zoologists in studies of the origin of particular species and groups

A crucial ecological question is the precise role of man in the recent history of the fauna. Knowledge of micro-evolution and adaptation during previous cold phases would provide a comparison for events of the last glaciation. The coincidence of the explosive evolution of *Homo sapiens* with the eclipse of a major portion of the fauna suggests a connection. The means by which depletion was achieved have not been established. It is essential that we know precisely what occurred if we are to explain man's own achievements in this brief period of time. Aspects of this extinction were discussed at a symposium during the VIIth Congress of the International Association for Quaternary Research. These papers were published (Martin and Wright, 1967) and are reviewed and discussed by Reed (1969).

The small mammals such as mice, shrews and voles, are rarely found in any numbers on prehistoric habitation sites. Their occasional presence is usually fortuitous. Small mammal bones are rare in the older collections from river gravels as the majority of bones were recovered by workmen. Nevertheless, careful excavation may produce a few. It is clear, however, that the majority of such deposits were neither conducive to the accumulation or preservation of remains of these small animals. In contrast a number of archaeological sites and geological deposits are particularly rich in the bones of small animals. These deposits may be summarised as follows:

1. Caves and shelters occupied as roosts by bats and birds.
2. Shafts and fissures.
3. Fine grained alluvial deposits in appropriate locations.
4. Preserved habitats such as building debris and turf stacks.

In these situations animal remains accumulate for the following reasons:

1. Death *in situ*.
2. Accumulation by carnivores.
3. Regurgitation by raptorial birds.
4. Natural pitfall traps.
5. Downwash.

Some bats roost and hibernate in caves, as may reptiles and amphibians; birds also roost there. In time the floor becomes littered with faeces and the carcasses of animals that died in the cave. Shafts and fissures are natural death traps for both large and small animals, depending on their size. Animals that have died elsewhere may also be washed into such structures. Gentle erosion may accumulate animal remains in fine grained deposits; if not re-sorted the bones are often well preserved. Occasionally the original habitat of the animal may be preserved. Remains of birds and small mammals are often found in and beneath tile and rubble falls in masonry buildings. The extensive use of turf as a building material in prehistory means that carcasses lying in the turf become incorporated in the structure. These may also be found on the original land surface preserved beneath prehistoric monuments. Carnivores frequently remove their prey from the kill site and take it to their den. Some, such as badgers, clean out their dens but others such as foxes do not. Raptorial birds (owls, hawks, etc) regurgitate in pellet form the indigestible bones and fur of their prey. This usually occurs at the roosting site so that there is an accumulation of pellets which in time will decay. Because of the diversity of roosting sites these pellets may be found almost anywhere.

Small mammals exist in a zone only a few inches above the surface where the climate is very different from that at higher levels. Whilst the tree cover influences the ground vegetation and the climate at the surface, different plant associations may produce very similar climatic conditions. Thus in terms of climate a species need not be confined to a single vegetational association. What does vary with the tree type is the amount of ground cover and the degree of protection from raptorial birds. Many raptors are unable to hunt in large tracts of forest despite an abundance of ground mammals.

There are many small mammals species in Europe abundantly present, and widely distributed. They occupy a great diversity of habitats. Because of this there is intense competition between species and any account of their ecology must take account of this competition and also predation. The abundance of related species in Europe makes them particularly significant in paleoecological studies. Unfortunately their ecology is poorly understood and until more work is done the potential of these species will not be realised.

The majority of general comments about the large mammals are equally applicable to the small ones. One important distinction must be made in respect of their evolution. Small mammals breed prolifically and rapidly and their generation time is much shorter than that of large mammals. As a result they can evolve faster. Thus studies of micro-evolution in small mammals may be of greater value in Pleistocene studies than similar work on large mammals. Unfortunately the best deposits of small mammals are of geological origin and are not associated with archaeological horizons.

It will be appreciated that much of the discussion in this chapter has centred on the secondary element of the ecosystem—the herbivores. Little attention has been paid to the tertiary element—the carnivores—of which, behaviourally, man is one. The reason is that the ecology of carnivores is directly related to their prey—the herbivores. Thus carnivores may transgress our typical zones. Many carnivores have a limited geographical range and these can be studied in the same way as other species that we have discussed.

Marine Mammals

The significance of marine mammals will depend on the proximity of the site to beaches, estuaries and rivers which these animals may visit. Some groups such as the eskimos have been particularly dependent on seals and walruses whilst other littoral peoples have extensively hunted whales, seals and related species. Marine mammals are frequently stranded on beaches and would have provided a valuable source of meat and oil. The history of whaling is quite well documented but its prehistory is still largely unexplored.

Reptiles, Amphibians and Fish

The bones of reptiles and amphibians are not common on excavations but they can be studied in the same way as small mammals. Remains of the bony fishes are valuable indications of the water regimes adjacent to habitation sites and the range of species exploited may indicate techniques employed in their capture. Further consideration of the site and species found may indicate the extent to which a group moved to collect the fish and might indicate seafaring ability.

Birds

Bird remains are of greater potential value in environmental studies than those of mammals. There are more species of birds than of mammals and their biological and ecological tolerances are frequently more specific. In the northern hemisphere birds have been more fully studied than any comparable animal group. Unfortunately there are few circumstances in antiquity that would result in the collection and preservation of a cross section of the avifauna. Almost all bird bones found on a site are there because man chose to gather these animals. Thus bird bones must almost invariably be interpreted in the light of human hunting and other activity. Because of the problem of sampling the ancient fauna it is inevitable that we shall be more concerned with individual species rather than faunas.

For a given species recorded on a site we are concerned to know of (a) the present and former distribution; (b) the apparent range of ecological tolerance and the factors which are critical in determining range, and (c) the seasonal distribution.

The geographical range of most avian species is wide. Thus of North American species with a limited range most cover a thousand miles north to south and east to west. The situation in Europe is similar, a visitor to southern or eastern Europe will find that the majority of birds seen are familiar even if their relative abundance is different. Therefore it is the ecological requirements rather than geographical distribution that concern the archaeologist. Some species occupy the same physical and ecological areas throughout the day, whilst others, notably at present starlings, pigeons and ducks move between feeding and roosting areas which are ecologically distinct. Knowledge of the species behaviour is thus important in deciding in what circumstances a species may have been captured.

For many years ornithologists have recognised the ecological dynamism of birds as they have adapted themselves to a rapidly changing habitat. We should remember that in a given time most bird species pass through more generations and experience a greater diversity of habitats than any mammal. We are accustomed by the ecological approach to ornithology to consider habitat groups—birds of the seashore, estuary, hedgerow, garden, downland, etc. Historically only a few of these habitats are applicable. The majority of lowland habitats consisted of mature woodland and the seral stages beneath this. Field, hedgerow, park and garden habitats are of more recent origin. The corollary of this is that the species occupying these newer niches have radiated to them from a limited number of primary habitats. In consequence the habits of species still found in primary habitats may also have changed and it is possible that some species are not members of the primary habitats but have moved into fill a vacant niche. Fortunately there still exist in Europe large tracts of land which have probably largely preserved the original habitat and its species. It is to such areas that we must turn for information about ecological tolerances of species in the past rather than to present studies in highly agricultural and urban landscapes such as Britain and Holland. These are in many cases also at the periphery of the Continental range of some species. With this evidence available and the probability of several species being present on a site where significant use was made of the avifauna, it should not be too difficult to indicate the general nature of the ground cover and the habitats available.

On non-subsistence sites, e.g. towns, villages and manor houses, the avifauna is likely to be somewhat different. At this stage in time we are dealing with a landscape with the secondary habitats noted above. Here we can expect to find examples of the birds that inhabit gardens, towns and buildings as well as a more exotic element. This exotic element will be recognised by the presence of birds that do not naturally belong in the domestic habitat. In the case of birds kept for ornament and pleasure these may well consist of the smaller brightly coloured songbirds as well as larger decorative species such as geese, peacocks and other foreign birds. A second element is likely to be birds used for sport. In particular the hawks and falcons were an important part of sporting activity in many parts of the world. A third element concerns birds associated with religious activity. Bird remains are associated with a number of ritual sites, for example the roman temples at Springhead in Kent and at St Albans in Hertfordshire. Elsewhere the association of birds with ritual activity is more dramatic as with the sacred Ibises from Egypt. These ritual occurrences are properly a study of the species in their ritual context rather than in relationship to the environment. They do, however, need to be distinguished from natural occurrences, sporting activities and subsistence collecting for food, feathers, etc.

The richness of the littoral and estuarine avifauna of the North Sea and Baltic basins in prehistoric times can now only be glimpsed at places like Foulness on the Essex coast. To the subsistence populations of the areas in prehistoric times these birds were of great importance. Seasonal activity and occupance of sites of this date has been regarded as normal. Bird remains can be used to deduce seasonal occupance of a site. Many birds are seasonal migrants and are to be found at a given spot only during part of the year. At some sites species are only found for a very short season in the course of their migration. Such species are referred to in that area as passage migrants. Thus the presence of these migratory species normally indicates that man was active in these areas at that time. The converse that man was absent from these areas except when the migrant species were there is not necessarily true. The seasonal influx of species in vast numbers would make them an attractive and readily available source of food. At other times of the year many of these species would be absent, and the population much smaller. It might then not be very worthwhile to try and catch them. In considering such a site we must take into account the total avifauna. This will probably consist of species resident throughout the year together with species present for a limited season. Such an occurrence does not prove seasonal occupance or hunting activity. It shows no more than that man took the species which were present. A more refined analysis of the remains is necessary to prove seasonal hunting activity and possibly occupance. A high proportion of the birds will be those born in the year. Knowledge of the growth of wildfowl will enable one to pinpoint the approximate age in months of the immature animals. Thus to prove seasonal activity it would be necessary to show that the immature element of the sample was of restricted age. If a sufficiently large sample is available one can then draw a strong inference that the adult species also represent a seasonal catch rather than a year-round one. The bone evidence can only demonstrate seasonality or otherwise in catching activity, it cannot prove seasonal occupance. Conversely it can show year-round exploitation and hence occupance.

On an archaeological time scale ornithological recording is of very recent origin. Although some species distributions and numbers can change dramatically in a few years others take much longer. Thus many changes now being observed had begun before ornithological recording and therefore we do not know when or where and in what circumstances these began. It is here that the archaeologist can make a valuable contribution to historical ornithology and ecology. The occurrence of a species at a time and place can in due course be fitted in with other similar observations extending the recording period far back in time. There are of course many hazards in using such data, negative information cannot be used as in modern observation but positive evidence can.

Although bird bones have been neglected in the past, and information from them wrongly interpreted it is no reflection of their intrinsic interest, only of our failure to systematise their study and apply ecological and cultural principles to their interpretation. Bird bones stand on a par with all other animal remains in their value to the archaeologist and ecologist.

The most recent general survey of the Pleistocene mammals of Europe is that of Kurtén (1968) and a comparable survey of the present fauna has been made by Corbet (1966). A convenient illustrated field guide to the birds of Europe is that by Peterson, Mountfort and Hollom (1954) and a similar volume is available for the mammals (Van den Brink, 1967). In North America the Pleistocene fauna is review by Hibbard et al. (1965).

The influence of the environment on faunas at a continental level is exemplified by the survey of Pleistocene biology and ecology in North America by Martin (1958) and Crisp (1959) has investigated recent changes in the environment and their effect on some plants and animals in Europe. The reciprocal relationship of archaeological and ecological studies is well illustrated in papers by Guilday and Parmalee (1965), Guilday, Hamilton and McCrady (1966) and Guilday, Martin and McCrady (1964). The contribution of archaeology to species history is shown in the studies of Guilday (1963), Reed (1965) and Degerbøl and Krog (1959). The fauna associated with ancient man has been investigated at a regional level by Clason (1967) in a valuable survey of faunal remains in Holland, and by Vereshchagin (1967) who considered the recent mammalian fauna of the Caucasus taking into account the archaeological evidence. The potential of the non-mammalian fauna is indicated by the studies of amphibian and reptilian faunas from Polish Pliocene and Pleistocene deposits by Mlynarski (1962), the survey of bird remains by Dawson (1966) and that of fish remains by Ryder (1969).

In interpreting the fossil assemblages it is necessary to consider its character and composition in the light of the dynamics of modern populations. A valuable theoretical discussion of population dynamics in regard to fossil assemblages is given by Craig and Oertel (1966).

References

Anderson, H. and Jorgenson, J. B. (1960). Decalcification of archaeological bones with histochemical interpretation of metachromasia. *Stain Technol.* **35**, 91–96.

Banfield, A. W. F. (1960). The use of caribou antler pedicels for age determination. *J. Wildl. Mgmt.* **24**, 99–102.

Boessneck, J. (1964). Über die osteologischen Arbeiten und Probleme des Tieranatomischen Instituts der Universität München. *Z. Agrargesch. Agrarsoziol.* **12** (2).

Boessneck, J., Müller, H. H., and Teichert, M. (1964). Osteologische Unterscheidungsmerkmale zwischen Schaf (*Ovis aries* Linne) und Ziege (*Capra hircus* Linne). *Kühn-Arch.* **78**, 1–129.

Brothwell, D. R. (1965). "Digging up Bones". The British Museum (Natural History), London.

Brothwell, D. and Brothwell, P. (1969). "Food in Antiquity." Thames and Hudson, London.

Chaplin, R. E. (1965). Animals in archaeology. *Antiquity*, **39**, 204–211.

Chaplin, R. E. (1966). "Reproduction in British Deer." Passmore Edwards Museum, London.

Chaplin, R. E. (1966). The animal remains from the Well Street Site, Coventry. *Trans. Bgham. Archaeol. Soc.* **81**, 130–138.

Chaplin, R. E. (1967). The character and function of osteological collections in provincial museums. *Mus. Assist. Group Newsletter*, (**9**), 5–10.

Chaplin, R.E. (1969). The use of non-morphological criteria in the study of animal domestication from bones found on archaeological sites. In "The Domestication and Exploitation of Plants and Animals". (Ucko and Dimbleby eds.) Duckworth, London.

Chaplin, R. E. and White, R. W. (1969). The use of tooth eruption and wear, body weight and antler characteristics in the age estimation of male wild and park fallow deer (*Dama dama*). *J. Zool. Lond.* **157**, 125–132.

Chaplin, R. E. and White, R. W. (1970). The sexual cycle and associated behaviour patterns in the fallow deer. *Deer*, **2**, 561–565.

Clark, J. G. D. (1965). "Prehistoric Europe. The Economic Basis." Methuen, London.

Clason, A. T. (1967). "Animal and Man in Holland's Past." J. B. Wolters, Groningen.

Corbet, G. B. (1964). "The Identification of British Mammals." British Museum, Natural History, London.

Corbet, G. B. (1966). "The Terrestrial Mammals of Western Europe." Foulis, London.

REFERENCES

Craig, G. Y. and Oertel, G. (1966). Deterministic models of living and fossil populations of animals. *Q. J. Geol. Soc. Lond.* **122**, 315–355.

Cranstone, B. A. L. (1969). Animal husbandry: the evidence from ethnography. *In* "The Domestication and Exploitation of Plants and Animals". (Ucko and Dimbleby eds.) pp. 247–263. Duckworth, London.

Crisp, D. J. (1959). The influence of climatic changes on animals and plants. *Geogrl. J.* **CXXV** (1), 1–19.

Darling, F. F. (1937). "A Herd of Red Deer." Oxford University Press, London.

Dawson, E. W. (1969). Bird remains in archaeology. *In* "Science in Archaeology" (Brothwell and Higgs eds.), pp. 359–369. Thames and Hudson, London.

Degerbøl, M. and Krog, H. (1959). "The Reindeer (*Rangifer tarandus* L.) in Denmark: Zoological and Geological Investigations of the Discoveries in Danish Pleistocene Deposits." Munksgaard, Copenhagen.

Duerst, J. (1926). Untersuchungmethoden am skelett bei Saugern. *In* Abderhaldens, "Handbuch der biologischen arbeitsmethoden," abt. VII, Methoden der vergleichenden morphologischen Forschung, Heft 2.

Empel, W. and Roskoz, T. (1963). Das Skelett der Gliedmassen des Wisents, *Bison bonasus* (Linnaeus, 1758). *Acta Theriologica* **VII**, 259–300.

Enlow, D. H. (1963). "Principles of Bone Remodelling." Charles C. Thomas, Springfield, Illinois.

Enlow, D. H. and Brown, S. O. (1956–58). A comparative histological study of fossil and recent bone tissues.
    Part 1. (1956), *Tex. J. Sci.* **VII** (4), 405–443.
    Part 2. (1957), *Tex. J. Sci.* **IX** (2), 186–214.
    Part 3. (1958), *Tex. J. Sci.* **X** (2), 187–230.

Evans, F. Gaynor (ed.), (1966). "Studies on the Anatomy and Function of Bone and Joints." Springer-Verlag, New York.

Ewbank *et al.* (1964). Sheep in the Iron Age: a method of study. *Proc. Prehist. Soc.* **30**, 423–426.

Garlick, J. D. (1969). Buried bone. *In* "Science in Archaeology" (Brothwell and Higgs eds.), pp. 503–512. Thames and Hudson, London.

Gilmore, R. M. (1946). To facilitate cooperation in the identification of mammal bones from archaeological sites. *Am. Antiq.* **12**, 49–50.

Gilmore, R. M. (1949). The identification and value of mammal bones from archaeological excavations. *J. Mammal.* **30**, 163–169.

Gooder, E., Woodfield, C. and Chaplin, R. (1966). The walls of Coventry. *Trans. Bgham. Archaeol. Soc.* **81**, 88–138.

Guilday, J. E. (1963). Pleistocene zoogeography of the lemming, *Dicrostonyx*. *Evolution*, **17**, (2), 194–197.

Guilday, J. E. (1963). The cup and pin game. *Penn. Archaeol.* **XXXIII** (4), 159–163.

Guilday, J. E., Martin, P. S. and McCrady, A. D. (1964), New Paris No. 4: A of Bootlegger Sink, York County, Pa. *Ann. Carnegie Mus.* **38**, 145–163.

Guilday, J. E., Martin, P. S. and McCrady, A.D. (1964). New Paris No. 4: A Pleistocene cave deposit in Bedford County, Pennsylvania. *Bull. natn. Speleological Soc.* **26**, 121–194.

Guilday, J. E. and Parmalee, P. W. (1965). Animal remains from the Sheep Rock Shelter (36 Hu 1), Huntingdon County, Pennsylvania. *Penn. Archaeol.* **XXXV** (1), 34–49.

Guilday, J. E., Parmalee, P. W. and Tanner, D. P. (1962). Aboriginal butchering techniques at the Eschelman site (36 LA 12). Lancaster County, Pennsylvania. *Penn. Archaeol.* **XXXII** (2), 59–83.

Habermehl, K. H. (1961). "Alterbestimmung bei Haustieren Pelztieren und bein jagdbaren wild." Paul Parey, Hamburg.

Hargrave, L. L. (1938). A plea for more careful preservation of all biological material from prehistoric sites. *Southwestern Lore*, December (1938), 47–51.

Hargrave, L. L. (1965). Identification of feather fragments by microstudies. *Am. Antiq.* **31**, 202–205.

Heizer, R. F. (1960). Physical examination of habitation residues. Viking Fund Publication: *Anthropology*, **28**, 93–157.

Hibbard, C. W., Ray, D. E., Savage, D. E., Taylor, D. W. and Guilday, J. E. (1965). Quaternary mammals of North America. In "The Quaternary of the United States" (Wright and Frey eds.), pp. 509–525. Princeton University Press.

Higham, C. F. W. (1969). The metrical attributes of two samples of bovine limb bones. *J. Zool. Lond.* **157**, 63–74.

Howard, M. M. (1963). The metrical determination of the metapodials and skulls of cattle. In "Man and Cattle" (Mourant and Levner eds.), Royal Anthropological Institute, Occasional Paper **18**, 91–100.

Jensen, B. and Nielsen, L. B. (1968). Age determination in the Red fox (*Vulpes vulpes* L.) from canine tooth sections. *Dan. Rev. Game Biol.* **5** (6), 15 pp.

Jubb, K. V. F. and Kennedy, P. C. (1963). "Pathology of Domestic Animals," Vol. 1. Academic Press, London.

Klevezal, G. A. and Kleinenberg, S. E. (1967). "Age Determination of Mammals by Layered Structure in Teeth and Bones." Moscow: Akademia Nauk SSSR, Institut Morphologii Zhivotnykh im A. N. Severtsova. Published by Izdatel Stvo "nauka". (Fisheries Research Board of Canada, translation series no. 1024 (1969).)

Kurtén, B. (1968). "Pleistocene Mammals of Europe." Weidenfeld and Nicolson, London.

Kyle, H. M. (1926). "The Biology of Fishes," London.

Low, W. A. and Cowan, I. McT. (1963). Age determination of deer by annular structure of dental cementum. *J. Wildl. Mgmt.* **27**, 466–471.

Lowe, V. P. W. (1967). Teeth as indicators of age with special reference to Red deer (*Cervus elaphus*) of known age from Rhum. *J. Zool. Lond.* **152**, 137–153.

Madsen, R. M. (1967). "Age determination of wildlife: a bibliography." Bibliography no. 2, U.S. Department of the Interior. Washington D.C.

Martin, P.S. (1958). Pleistocene ecology and biogeography of North America. In "Zoogeography". (Hubbs, C. L. ed.). American Association for the Advancement of Science, Publication 51, pp. 375–420.

Martin, P.S. and Wright, H. E. (eds) (1967). "Pleistocene Extinctions: The Search for a Cause." Yale University Press, New Haven.

McDiarmid, A. (1966). Safety precautions at post mortem examinations. *Bull. Mammal Soc.* No. 26, p. 17.

Messinger, N. G. (1965). Methods used for the identification of feather remains from Wetherill Mesa. *Am. Antiq.* **31**, 206–215.

REFERENCES

Mlynarski, M. (1962). Notes on the amphibian and reptilian fauna of the Polish Pliocene and Early Pleistocene. *Acta Zoologica Cracoviensia.* **VII**, (11), 177–193.

Nishiwaki, M., Hibiya, T. and Ohsumi, S. (1958). Age study of sperm whale based on reading of tooth laminations. *Scient Rept. Whales Res. Inst.,* Tokyo No. 13, 135–150.

Parmalee, P. W. (1959). Use of mammalian skulls and mandibles by prehistoric indians of Illinois. *Trans. Ill. St. Acad. Sci.* **52**, 85–95.

Parmalee, P. W. (1965). The food economy of archaic and woodland peoples at the Tick Creek site, Missouri. *Missouri Archaeologist,* **27**, 1–34.

Peterson, R., Mountfort, G. and Hollom, P. (1954). "A Field Guide to the Birds of Britain and Europe." Collins, London.

Quimby, D. C. and Gaab, J. E. (1957). Mandibular dentition as an age indicator in Rocky Mountain Elk. *J. Wildl. Mgmt.* **21**, 435–451.

Reed, C. A. (1963). Osteo-archaeology. *In* "Science in Archaeology" (Brothwell and Higgs eds.) 1st edition, Thames and Hudson, London.

Reed, C. A. (1965). Imperial Sassanian hunting of pig and fallow deer and problems of survival of these animals today in Iran. *Postilla,* **92**, 1–23.

Reed, C. A. (1969). They never found the ark. *Ecology,* **50**, 343–346.

Rodahl, K., Nicholson, J. T. and Brown, E. M. (eds) (1960). "Bone as a Tissue." McGraw-Hill Book Company, New York.

Robinette, W. L., Jones, D. A., Rogers, G. and Gashwiler, J. S. (1957). Notes on tooth development and wear for Rocky Mountain Mule Deer. *J. Wildl. Mgmt.* **21**, 134–153.

Ryder, M. L. (1966). The exploitation of animals by man. *Adv. Sci.* **23**, 9–18.

Ryder, M. L. (1969). Remains of fishes and other aquatic animals. *In* "Science in Archaeology" (Brothwell and Higgs, eds), pp. 376–394. Thames and Hudson, London.

Ryder, M. L. (1969). Changes in the fleece of sheep following domestication (with a note on the coat of cattle). *In* "The Domestication and Exploitation of Plants and Animals," (Ucko and Dimbleby eds), pp. 495–521. Duckworth, London.

Schmid, E. (1967). Tierreste aus einer Grossküche von Augusta Raurica. *Basler Stadtbuch,* 176–186.

Scott, J. H. and Symons, N. B. B. (1958). "Introduction to Dental Anatomy." E. and S. Livingstone, Ltd., Edinburgh and London.

Sergeant, D. (1967). Age determination of land mammals from annuli. *Z. Saügetierk.* **32** (5), 297–300.

Silver, I. A. (1969). The ageing of domestic animals. *In* "Science in Archaeology," (Brothwell and Higgs eds). Thames and Hudson, London.

Simpson, G., Roe, A. and Lewontin, R. (1960). "Quantitative Zoology." 2nd edition, Harcourt Brace and World Inc., New York.

Skinner, M. F. and Kaisen, O. C. (1947). The fossil bison of Alaska and a preliminary revision of the genus. *Bull. Am. Mus. nat. Hist.* **89**, 127–256.

Treganza, A. F. and Cook, S. F. (1948). The quantitative investigation of aboriginal sites: complete excavation with physical and archaeological analysis of a single mound. *Am. Antiq.* **13**, 287–297.

Trow-Smith, R. (1957). "A History of British Livestock Husbandry to 1700." Routledge and Kegan Paul, London.

Trow-Smith, R. (1959). "A History of British Livestock Husbandry 1700–1900." Routledge and Kegan Paul, London.
Van den Brink, F. H. (1967). "A Field Guide to the Mammals of Britain and Europe." Collins, London.
Vereshchagin, N. K. (1967). "The Mammals of the Caucasus: a History of the Evolution of the Fauna." Translated from the Russian by Israel Program for Scientific Translation.
Wells, A. L. (1958). "The Observer's Book of Sea Fishes." Warne, London.
White, T. E. (1952–55). Observations on the butchering technique of some aboriginal peoples. *Am. Antiq.* **17**, 337–338: **19**, 160–164: **19**, 254–264: **21**, 170–178.
White, T. E. (1956). The study of osteological materials in the Plains. *Am. Antiq.* **21**, 401–404.
Wolstenholme, G. E. W. and O'Connor eds. (1956). "Ciba Foundation Symposium on Bone Structure and Metabolism." J. and A. Churchill Ltd., London.
Zalkin, V. I. (1960). Die Veränderlichkeit der Metapodien und ihre Bedeutung für die Erforschung des groben Hornviehs der Frühgeschichte. *Bjull. Mosk. Obsh. Isp. Prir, Biol. N.S.* **65**, 109–126.
Zalkin, V. I. (1961). The variability of metapodialia in sheep. *Bjull. Mosk. Obsh. Isp. Prir. Otd. Biol.* **66**, 115–132.

Index

A

Actinomycosis, 113
Adaptation, to environment,
  ecological, 151
  environmental, 151
  physiological, 151
Age determination, 76–90
  by antler development, 77, 89
  from bones, 76–90
  by epiphyseal fusion, 9, 77, 80–81
  by incremental structures, 77, 84, 85
  by size and form, 77, 89
  by suture closure, 77, 81–84
  techniques used, 76, 77 et seq.
  by tooth eruption, 77, 78–80
  by tooth wear, 77, 85, 86–88
  by qualitative features, 77, 90
Alpine Zone, 145, 146, 149
Amphibians, 48, 153, 155
Animal remains,
  and dating, 143
  examinations and studies available, 61
  and past environments, 143–159
Antler development, 89

B

Birds, 38, 45, 46, 48, 124, 153, 155–159
  distributions of, 156
  geographical range, 156
  habitats, 156
  history of, 156
Bleaching, of bone, 51–53
Boiling bones, 51
Bone,
  age estimation, 9 (see also Age determination)
  Blood grouping from, 28
  cancellous, 5
  chemical composition of, 12, 13
  chemical properties of, 12, 13
  commercially discarded, 14
  development, 7–11
  discarded, 15
  formation of, 7–9
  fracture and repair of, 11
  function of, 1–7
  growth of, 7–11
  hydroxyapatite in, 13
  infection of, 11, 108–119
  marrow, 2, 5, 6, 8
  mineral of, 13
  pathology of, 108–119
  physical properties of, 12, 13
  radiographs of, 5, 22, 81, 113
  as raw material, 123
  structure of, 1–19
  types of, 2
Bones,
  age determination from, 76–90
  biological properties of, 1–19
  of birds, 38, 45, 46, 48, 124, 153, 155–159
  bleaching of, 51–53
  butchery of, 66, 67, 123, 133–138
  cartilage bones, 7
  of Cetacea, 46
  cleaning of, 26, 27, 28
  collagen in, 13
  compact bones, 2, 3, 6
  comparative collections of, 41–49
  contamination of, 2, 5, 28, 120, 121
  cooking of, 14, 18, 64
  dating from, 27
  degreasing of, 51–52
  destruction of, 13–19
  diseased, 1, 2, 11, 109, 112, 113 et seq.
  distortion of, 28
  disposal of, 13–19, 120–122
  excavation of, 20, 21, 24–31
  of fish, 25, 45, 48, 155
  flattened bones, 2, 3, 4
  identification of, 27–54

Bones—*cont.*
  labelling of, 31–35
  long bones, 2, 3, 4, 5
  measurement of, 91–99
  membrane bones, 7, 8
  as organs, 2
  ossification of, 8, 9
  packaging of, 31–35
  pathology of, 109–119
  photography of, 2, 27
  as playthings, 1, 2, 3
  preparation of, 50–54
  preservation of, in soil, 13–19
  preservation of excavated bone, 24–36
  processing of, 20–36
  recording of, 20–36
  reference collections of, 41–49
  repair of, 24, 35
  of reptiles, 48, 53, 155
  sex determination from, 91, 92, 100–107
  sexual dimorphism in, 39, 92, 93, 102, 103–7, 112, 140, 141
  size indices from, 91, 92
  survival of, 16–19
  as tools, 123
  transport of, 31
  types of, 2
  washing of, 28
Bone evidence,
  of butchery practice, 66, 67, 123, 133, 138
  economic, 79, 120, 121, 123
  of environment, 144–159
  of food supply, 133–138
  of hunting practice, 56
  of husbandry practice, 76, 79, 135–138
  interpretation of, 120–142
  from refuse, 120–121
  of selective killing, 135
  of trade, 135
Bone pathology, 109–119
Bone studies
  abroad, 20–23
  determining age at death, 128, 76–90
  establishing economic data, 123
  on excavations, 20, 21
  field work in, 23
  finance of, 21, 22
  and food economy, 123, 133–135
  history of, 20
  and integrated studies, 122, 123
  organisation of, 21
  place in archaeology, 20–24
  preliminary tasks in, 35
  quantifying species in, 63–70
    fragments method, 64–67
    minimum numbers method, 69
    weight method, 67–69
  reburial of material in, 24
  repair work, 24, 35
  requirements for collection of bones, 24
  role of excavators in, 22
  role of specialists in, 21, 23
  specialists in, 20, 21
  writers on, 20
  and zoological research, 23
Boreal Forest, 145, 146, 148
Butchery practices, 66, 67, 123, 133, 134, 135–138

C

Callipers, 95–97
Carbon 14 dating, 27
Carcasses,
  preparation of, 51–54
  sources of, 42
Cartilage, 5, 7, 8, 9, 80
Castration, 112
  effects of, 81
Cementum annuli, 84, 85
  ageing from, 84
  nature of, 84
  staining of, 85
Chondrodystrophia foetalis, 109
Cleaning of bone, 26, 27, 28
Collection of bone,
  comparative collections, 41–49
  handling, 24
  requirements for, 24
  reference collections, 41–49
Commercial preparation of bone, 51
Compact bones, 2, 3, 5, 6
  functions of, 2
  structure of, 6
Comparative collections, 41–49
  access to, 41, 42
  costs of, 41, 42
  establishment of, 43 *et seq.*
  general collections, 42
  museum collections, 41, 42
  requirements of, 41
  sources of specimens for, 42
  special collections, 49, 50
  species required for, 42, 45–48
Complete burial, 65
  of dog, 65

Congenital hip dysplasia, 114
Coniferous forest (see Boreal forest)
Contamination of bone, 25, 28, 120, 121

D

Dating, from bone, 27
    carbon 14, 27
    use of species in, 143
Deciduous forest, 145, 146, 147, 148
Degenerative arthropathy, 114
Degreasing of bone, 51, 52
Dermestes, use of in bone preparation, 53
Determination of age, 76–90
Determination of sex,
    from bones, 91, 92, 100–107
Disease in bones, 1, 2, 11, 109, 112, 113 et seq.
    classification of, 108, 109
Dyschondroplasia, 109

E

Ecology, 144, 145, 151, 152, 153, 154
    (*see* Adaptations, Environment, Faunas, Habitats)
Endoskeleton, 1
Environment,
    adaptations to, 147, 148, 151
    animals as indicators of, 143–159
    changes in, 151
    variations in, 145–149
Enzymes,
    use of in preparation of bones, 53
Epiphyseal fusion, 17, 80–81
Evolution,
    of faunas, 143–149
    of species, 147, 148, 151, 152
Excavated bones,
    cleaning of, 26, 27, 28
    washing of, 28
Excavations, 20, 21
    planning of, 20
Exoskeleton, 1

F

Faunas, 143–149
    adaptations of, 146, 147
    of Alpine zone, 145, 146, 147, 149
    of Boreal forest, 145, 146, 147, 148
    British, 144
    changes in, 144, 145
    of Deciduous forest, 145, 146, 147
    European, 144, 145
    of Steppe zone, 145, 146, 147, 148
    terrestrial, 143
    of transitional zones, 145, 146, 147
        amphibian, 155, 159
        avian, 154, 155, 156, 157, 158, 159
        mammalian, 147, 148, 153, 154
        piscian, 155, 159
        reptilian, 155, 159
    tundra, 145, 146, 148, 149
Fish, 25, 45, 48, 155, 159
Flattened bones, 2, 3, 4
    longitudinal section of, 4
    mechanical function, 2
    protective function, 2
Fragments method, of quantifying species, 64–67

G

Growth of bone, 7–12, 81, 89
    appositional, 8
    longitudinal, 9

H

Habitats, 145–159
    adaptations to, 146–148, 151
    species distribution in, 146
    zonation of, 145, 146, 151
Horn, 139 et seq.
Horncore, 139, 140, 141
Hyperthyroidism, 112
Hypertrophic osteopathy, 113
Hypopituitary dwarfism, 112
Hypothyroidism, 112

I

Identification of bone, 37–54
    problems of, 37–41
Infective arthritis, 117
Insect remains, 26, 30, 31
Interpretation of bone evidence, 120–142
    ancillary evidence, use of, 122, 123
    limiting factors in, 120–122
    site examples, 123–142

J

Joint disease, 114–119
    classification of, 114

K

Kalmyk cattle, 103

L

Labelling of bones, 31–35
Lesions, 108, 109, 116–119
  classification of, 108
Longbones, 2, 3, 4, 8, 9
  longitudinal section of, 2, 4
  mechanical function of, 2, 5
  physiological function of, 6
  structural function of, 6

M

Maceration,
  use in preparation of bones, 51, 53
Mammals,
  adaptations in, 147, 148, 153, 154
  distributions of, 144
  faunas, 147, 148, 153, 154
Marine mammals, 155
Measurement of bone,
  accuracy, 98, 99
  code of, 94
  collection of, 94
  errors in, 92, 99
  equipment for, 95–98
  height coefficients, 92
  instruments for, 95–98
  in large animals, 75
  publication of, 93, 94
  for sex determination, 100
  and sexual dimorphism, 92, 93, 98
  and size, 91, 93, 98
  in small animals, 95
  specification of, 94
  standardisation of, 94
  technique, 98, 99
  use of, 91–99
  weight coefficients, 92
Medieval workshop debris, 138 *et seq.*
Membrane bones,
  formation of, 8
Microscopes,
  low power, 98
  for measurement, 98
  travelling, 98
Middens, 120, 121
Minimum numbers, 69
  determination of, 70
  use of, 69
Muscles, 2, 11

O

Ossification, 8, 9
  centres, 9
  endochondral, 8
  intramembranous, 8

Osteitis, 113
Osteo-archaeology, 23, 40, 50
Osteoblast cells, 8
Osteoclast cells, 8
Osteodystrophia fibrosa, 110, 111
Osteogenesis, 7, 8
Osteological collections, 41
  in museums, 41
  specialised, 49
  verified, 43
Osteomalacia, 110, 111
Osteometric equipment, 95, 96
  callipers, 95–97
  microscope, 98
  osteometric board, 96, 97
Osteomyelitis, 113
Osteoporosis, 110
Osteosis, 112

P

Packaging of bone, 31–35
  deterioration of materials, 33
Palaeobotany, 62
Palaeozoology, 62
Parasites, 27, 30, 31
Pathological conditions,
  in bone, 108–119
  classification of, 109
  origin and causes of, 109–117
  types of, 109–117
Perichondrium, 8
Periosteum, 6, 9, 11
  infection of, 113
Pleistocene,
  chronology of, 143
  faunas of, 143, 144
  studies, 155, 159
Polydactylia, 110
Polyvinyl acetate, 127
Precautions,
  in preparation of bone, 51, 54
Prehistoric sites, 26
Preparation of bones, 50–54
  equipment for, 51, 52
  methods of, 51, 52
  precautions in, 51, 54
Preservation of excavated bones, 24–36
Preservation of bone in soil (*see* Survival of buried bone)

Q

Quantification of species, 63–70
  by fragments, 64–67

INDEX

Quantification of species—*cont.*
  methods of, 63, 64
  by minimum numbers, 69–70
  by weight, 67–69

R

Radiographs of bone, 5, 22, 81, 113
Reference collections (*see* comparative collections)
  establishment of, 37–54
Refuse,
  accumulation of, 13–19, 121, 122
  commercial, 13–19, 121, 122
  dispersal of, 13–19, 121
  domestic, 13–19, 121
  evidence from, 120, 121
  middens, 13–19, 120, 121
  nature of, 13–19, 120, 121
  origin of, 13–19, 121
  sources, 13–19, 121
  use of, 120, 121
  as zoological evidence, 121
Repair, of bone, 24, 25
Reptiles, 48, 155
Requirements for collection of bone, 24
Rickets, 111
Ringbone, 117
Roman bone, 16, 18, 56, 57
Romano–British sites, 44
Romanov sheep, 92

S

Sampling, 25–26
Saxon bones, 102
Saxon farm Whitehall,
  study of, 124
Sex,
  determination of, 91, 92, 100–107
Sexual dimorphism, 39, 92, 93, 102, 103–107, 112, 140, 141
  effect of castration, 101
Sieving, 25, 26
Sites,
  post Glacial, 144, 145, 151
  post Medieval, 47
  post-Neolithic, 43
  post-Pleistocene, 43
  post-Roman, 43
  Mesolithic, 43, 44, 47
  Neolithic, 44
  Prehistoric, 26
  Roman, 56, 57
  Romano–British, 44
  Types of, 153

Site studies, 55–75, 120–142
  ancillary data in, 61
  cultural context of, 55–58
  examinations carried out in, 58–63
  general planning of, 55
  locational studies in, 58
  minimum evidence requirements in, 59–61
  phase studies in, 57, 58
  physical contexts, 56
  quantification in, 63, 64
  specification of examinations in, 62
  topic definition, 58
  working procedure in, 63
  working units in, 57
Skeleton, 1–7, 25, 26, 27, 29, 37, 38–40, 42, 53, 77
  endoskeleton, 1
  excavation of, 29
  exoskeleton, 1
  internal hard, 1
  lifting of, 26, 27, 29, 30
  mechanical functions of, 1, 2
  physiological functions of, 1, 2
  structural functions of, 1, 2
  structure of, 1–7
  transport of, 31–34
  treatment of, 29–34
  photography of, 26, 27, 29, 30
Sodium perborate, 51, 52
Soil analysis, 30
Soil conditions, 16, 17
Spavin, 117
Specialists, 20, 21–24 *et seq.*
  abroad, 20, 22, 23, 24
  tasks of, 23
  work of, 23 *et seq.*
Steppe zone, 145, 146, 148
Stone age material, 143
Structure of bone, 1–19
Survival of buried bone,
  by deep freezing, 16
  degradation processes, effect of, 15, 16
  by dessication, 16
  differential preservation, 16, 18
  factors affecting, 13–19
  frost heave, effect of, 18
  by mummification, 16
  in oilfield seepages, 16
  in peat bogs, 16
  pH, effect of, 15, 17, 18
  plant roots, effect of, 17
  scavengers, effect of, 14
    dogs, 14
  soil bacteria, effect of, 16

Survival of buried bone—*cont.*
  soil characteristics, effect of, 16, 17
  soil creep, effect of, 18
  soil structure, effect of, 16, 17
  soil water regime, effect of, 16, 17
  solifluxion, effect of, 18
  weather, effect of, 15, 18
Syndactylia, 110

T

Teeth, 12
  age determination from, 12, 77–88
  cementum annuli in, 84
  eruption of, 77, 78–80, 85, 88
  preservation of,
  wear, 77, 78–80, 85, 88
Transitional zones, 145, 147
Transportation of bone, 31–35
Tumours in bone tissue, 113
Tundra zone, 145, 146, 148

V

Vacuum impregnation, 28, 35
Vegetation zones,
  alpine, 145, 146, 149
  Boreal forest, 145, 146, 148
  Deciduous forest, 145, 146, 147, 148
  Steppe, 145, 146, 148
  transitional, 145, 147
  Tundra, 145, 146, 148
Vertebrae, 2, 3, 4
  mechanical function of, 2

W

Washing of bone, 28
Weight method of quantifying species, 67–69
Wroxeter, Roman town, 122